Ham Radio Horizons: The Book

By Peter R. O'Dell

Published by

CQ Communications Inc.
76 North Broadway
Hicksville, NY 11801

The American Radio
Relay League, Inc.
Newington, CT 06111

Jointly Published By

CQ Communications, Inc.
76 North Broadway
Hicksville, NY 11801

American Radio Relay League, Inc.
225 Main Street
Newington, CT 06111

Library of Congress Catalog Card Number: 92-073952

ISBN: 0-943016-03-7

For Sam, Jose, and Willie

Thanks for giving me the
courage to dream, to search,
to act, and to enjoy.

Contents

Preface

Have you every wondered about ham radio? Who the hams are? What they do? And how do they do it? This book answers those questions and more.

Although ham radio has its roots in the first forms of electronic communications, it is as modern and forward looking as the most sophisticated computer network. Yesterday's ham wound coils on Quaker Oats boxes and hand-wired almost every piece of equipment in his "radio shack." Today's ham buys almost all his equipment ready-built and spends time interfacing his high-tech station to his computer. The props have changed, but the spirit is the same.

You can be a part of it. You might want to talk to friends across town. Or you might want to keep in touch with your brother on the other side of the country. Perhaps you want to talk to an astronaut circling the earth. Ham radio is all these things and more. Above all, it is two-way person-to-person communications. Ham radio provides us with a means to participate and communicate with other citizens of our global village. It is the best way to make friends around the world without respect to age, gender, religion, political persuasion, or any of the other "categories" that tend to divide us.

It has never been easier to become a ham. Test preparation material abounds, and there is even a way to become a ham without learning Morse code! Equipment is plentiful and comparatively inexpensive. Why hold back?

1

Amateur Radio —Fun For Everyone

AN HISTORIC MOMENT

Jim Steffen sat on edge of his seat. He and four colleagues had their eyes glued to the clock. A scratchy buzz of static came from the radio receiver. As the anticipated minute approached, the static began to give way to a hissy, crackly sound implying a weak human voice. The sound slowly became clearer. Steffen looked at his cohorts and grinned in eager anticipation as they sat with hands poised over a variety of buttons and knobs. As the noise from the radio got louder, they heard it:

"...Hawaii. Now have continental US in sight, approaching West Coast." It was the voice of US space shuttle pilot Lt Colonel Ken Cameron, orbiting aboard *Atlantis*, 400 miles above the Pacific Ocean. "Everything checks out, we're go to receive video."

At that, Steffen threw a switch and pressed the playback button on the VCR. The group anxiously scanned an assortment of meters, making minor adjustments as they watched their video shuttle graphic dissolve to a shot of Steffen waving and smiling.

"I've got it now," exclaimed Cameron. "It looks black and white, pretty clear, not too much snow."

A group of ham radio operators from California beamed this television picture to astronauts aboard the Space Shuttle Atlantis. *Shuttle pilot Ken Cameron, whose Amateur Radio call sign is KB5AWP, was excited to receive the first-ever "fast-scan TV" sent to an orbiting spacecraft. He radioed down: "Jim, it was great!"* (photo by Jim Steffen)

Another TV picture sent up to Atlantis *showed one of the Amateur Radio operators giving the Vulcan "live long and prosper" salute from Star Trek.*

Jim and his friends let out a whoop! They'd done it—successfully transmitted the world's first live-action television to a manned spacecraft!

As the tape ended with Jim giving the *Star Trek* Vulcan salute and saying, "Live long and prosper," Cameron's voice came to them again.

"Jim, it was great!" reported the astronaut, "That is a first! We got the whole tape, then I saw you pick up the mike and talk to me. I had varying qualities. I couldn't get much color up here, but a very good quality picture of you and your compatriots."

Jim Steffen's elated technical crew wasn't stationed in a NASA ground station or government laboratory. Using Amateur Radio technology, they had made history in Jim's Long Beach, California, living room.

FIRE THREATENS YOSEMITE

The rangers at Yosemite National Park were concerned. The fire had raged through part of the Stanislaus National Forest and was threatening the towns of El Portal and Yosemite West. Forest Supervisor Jan Wold cast a worried glance at the firefighter command crew in the Park Communications Center in Yosemite Valley.

"Still no telephone circuits to Sonora," sighed Assistant Fire Supervisor Gary Biehl. "They say it'll be two to three days before the portable cellular phone gear can be brought in from National Headquarters in Boise."

Wold grimaced. The wind had just shifted. They'd have their best shot at stopping the holocaust before it reached the nearby communities and the valuable trees of Merced Grove if they could only move their fire teams a couple of miles. Unfortunately, the craggy, uneven terrain had made it impossible to contact the team leaders by two-way radio.

As fire threatened Yosemite National Park, ham radio operators helped deploy firefighters and equipment. Whenever an emergency threatens, ham radio operators are on the scene to provide communications support for local agencies.

HAM RADIO IN A NUTSHELL

It's not easy to say what Amateur Radio is all about in a few words. To give you some ideas, answer the following questions. Have you ever . . .

• Saved someone's life by talking?
• Visited a close friend in a foreign country?
• Had a pleasant, noise-free conversation with several friends while driving to work—alone?
• Sat in your living room and had a chat with an astronaut orbiting the earth in a space shuttle?
• Used your computer to send a holiday greeting from your Aunt Sarah across town to your cousin Marvin at college on the Coast?
• Held a conversation from your house to a stranger's house in Europe via an orbiting satellite?
• Called home from your car to speak to a member of your family, without using an expensive cellular telephone?
• Taken part in a sporting event "live" with other participants from every US state and more than 200 countries?
• Helped local authorities watch over the safety of spectators at a parade, marathon or bicycle race?
• Built a two-way radio from scratch and then used it to talk to friendly people on other continents?
• Written a computer program that lets you send data files and messages to friends across town without tying up your telephone?

The world's Amateur Radio operators, or hams, as they're commonly called, can answer an emphatic **yes** to all of the above. Ham radio is a communications service that allows nearly everyone to enjoy a unique and always-exciting hobby. Hams can use just about any means of communications you can imagine: voice, Morse code, computer-to-computer networking, television and satellites, to name just a few.

Hams are proud to be part of a global fraternity, one that promises excitement and adventure whenever the rig is fired up, anywhere around the world.

Then a hand-held radio on Fred White's belt came to life. "The portable repeater is up at Fire Camp Mather. Command Post, do you read me? Over."

White held the radio to his mouth and replied, "Roger, Mather Camp. This is Command Post. Solid copy."

"Okay," came the reply, "I have Team Red's squad leader here awaiting instructions on where to deploy."

"Good work. I'll put Gary on now so he can talk direct."

As White handed the radio to Biehl, the Assistant Fire Supervisor's frown turned to relief. He began to issue orders and the fire line teams moved into position just in time to cut off the blaze from its path toward Merced Grove, El Portal and Foresta.

The crack emergency communications support team that set up the vital backup system was not a National Guard unit or AT&T repair crew. It was a band of 100 volunteer Amateur Radio operators, some of whom had driven up to several hundred miles to install specially designed homemade radio relay equipment and power generators. The damage to Yosemite was significant, but all agreed it would have been far worse if the emergency communications hadn't been available.

"WE HAVE NO WEAPONS..."———————

Pentagon Intelligence officials read the latest dispatches. From directly inside Kuwait City, a single station was transmitting first-person accounts of troop movements and bombardment reports. Computer terminals in stations on the US mainland, in Guatemala, Lebanon, Sweden, The Netherlands, Egypt and Switzerland, and aboard the carrier *USS John F. Kennedy* passed along screens full of text describing the gruesome action. The reports were flashed to United Nations military intelligence units and faxed to the US State Department for analysis. Situation updates were encoded and dispatched to officers commanding the vast UN army as it moved across the borders into Kuwait and Iraq.

Operators stood duty around the clock, monitoring the sporadic transmissions, wondering whether the previous message had been the last. They knew the station in Kuwait City could be shut down at any moment, its operator executed and its equipment destroyed. The radio operator himself stated the Iraqis considered his transmitting equipment "as more than weapons." Yet, the sporadic bulletins continued to arrive, the station's powerful signal unavoidably serving as a beacon for radio direction-finding equipment. All the Iraqis would need was a location fix and a single missile or small force of soldiers could end the communications lifeline.

The tension eased as the swift military maneuver concluded in a matter of days, and the routed Iraqis retreated from Kuwait. Jubilant UN troops entered the liberated city and officials hastened to congratulate the lone radio operator.

The brave soul who'd been the last free voice from Kuwait amid the scene of terror and demolition was a civilian Amateur Radio operator. Setting aside concerns for his personal safety, Abdul Jabbar Marafi had kept the UN forces and listeners around the world apprised of the situation around him. His station had been the only means of dependable communications out of Kuwait City.

SOVIET COUP

In August 1991, hardline Communists in the Soviet military were opposed to President Mikhail Gorbachev's sweeping reforms and his push toward Soviet democracy. There was little warning, however, when the country's new direction was threatened by a single incident.

On the morning of August 18, Moscow TV showed a British movie. The radio played classical music. There was no news to be found. Yet, something critical was happening.

The telephone rang in a Moscow apartment. Close advisors to Boris Yeltsin, president of the Russian Republic, notified some communications hobbyists that their help was needed urgently. They were told to bring

WHERE DOES HAM RADIO FIT IN?

Amateur Radio is just one of many types of communications services to use the airwaves. Others include:

- the AM, FM and television broadcasters we're all familiar with
- police, fire, ambulance and other public-safety groups
- tow trucks and taxis
- ships and airplanes
- communications satellites and manned spacecraft
- cordless and cellular telephones
- nursery monitors
- Citizens Band (CB) radio
- shortwave broadcasting
- radar

and more.

It's not hard to see that all these transmissions would interfere with each other if they all were on the same place on a radio dial. That's one of the main reasons for regulating those who use the airwaves. The FCC, an agency of the US government, tells each type of service which frequencies it can use. The FCC also requires licenses for many of these activities. These licenses authorize operation in a particular place on the airwaves.

Just as sound ranges from the low-pitched rumble of thunder to high-pitched trill of a songbird, radio signals cover a wide spectrum, from low to high frequencies.

---

some equipment to Yeltsin's headquarters in the Russian Parliament, the "Russian White House," immediately.

There, a dozen Soviet citizens and one visiting American found themselves in the thick of a military coup designed to oust Gorbachev. Yeltsin was determined to resist. When the radio equipment was brought in and set up, Yeltsin and other top-ranking officials began round-the-clock broadcasts. They sent out assurances of solidarity to rally the people against the coup, and issued news and information to the waiting world.

As coup leaders attempted to jam the transmissions, the radio operators used the tricks and skills they had honed in years of operating to elude the jammers. Yeltsin's powerful signals went out on broadcast, short-wave and Amateur Radio frequencies. Anxious listeners around the world hung on the words coming from the meager station. At the same time, radio and TV stations and newspapers disseminated the contents of the clandestine broadcasts.

By the next week, it was over. The coup attempt had been aborted, its instigators captured or expelled.

The radio equipment had been installed and operated entirely by volunteer Amateur Radio operators.

WHAT IS AMATEUR RADIO?

These stories are true, and not as unusual as you might imagine. Since the earliest days of wireless communications, Amateur Radio operators have saved thousands of lives, prevented billions of dollars in property damage and risked—even given—their lives.

Ham radio operators are there when information needs to get through. Their insatiable curiosity has led to the technology that has given us the satellite, television, broadcasting, telephone, computer and space communications we often take for granted.

When you get your ham license, you'll be in good company. There are more than half a million Amateur Radio licensees in the US and millions more worldwide. Walter Cronkite is a ham, and so are singer Ronnie Milsap and Jordan's King Hussein. Former US Senator Barry Goldwater, who ran for President in 1964, is a ham. So are Dick Rutan and Jeana Yeager, first to pilot an ultralight aircraft nonstop around the world. Chances are, you have relatives, friends or neighbors who are hams.

CB, Cellular and Ham: A Brief Comparison

Ham radio is unique. Hams can talk to other hams on the other side of the world or on the other side of town.

Ham radio operators are there when information needs to get through.

TECHNOLOGY: HAMS LEAD THE WAY

When it comes time to get ham radio equipment, you can buy everything you'll need off the shelf (or through the mail). If you prefer, you also can design and build it yourself.

The FCC permits—even encourages—hams to design build, repair and modify their own radio equipment. The original "wireless" pioneers who developed the foundations upon which most of our modern radio communications systems are built, were amateurs. To continue this rich tradition, the FCC expects and encourages radio amateurs to experiment, tinker and create new ways of communicating.

There were radio amateurs even before there was radio. They didn't have licenses or call signs (after all, there was no reason to bother yet). These early experimenters had many of the same qualities as today's hams: intelligence, curiosity, creativity and a fascination with the art and science of radio communications. The first successful two-way radio contact was made by Guglielmo Marconi, a self-described radio amateur.

Most of today's hams operate store-bought equipment. You'd be hard-pressed to find an experienced ham who has never built an antenna or other piece of gear, however. The cutting edge of technology gets sharper every day; it takes special expertise to stay on top of all the latest breakthroughs. But inventive hams continue to push the limits in these areas, to name just a few:

- *Amateur Radio satellites*, designed and built by hams, send rich visual images to earth, and do so at a fraction of the cost of comparable commercial spacecraft.
- *Computer-to-computer hookups* via radio are polished and refined as software wizards try out new programs and equipment to enhance data-transfer systems.
- Hams have been *bouncing signals* to each other *off the moon* since the early 1950s—before the first man-made satellites were launched.

You don't need an advanced electronics degree to continue this rich tradition. Hams who enjoy nothing more than operating their store-bought radios are constantly devising ways of improving the efficiency of their stations. Whether you're an electronics whiz, a grandmother or a 6th grader, you can find a way to contribute to the state of the Amateur Radio art.

Hams can operate their radios from their cars or from space. Hams can travel the world without a passport—and all that luggage.

You can talk to people on a CB radio, but your range is normally limited to a few miles. CBers can use only 40 channels.

Ham radio operators have many *thousands* of times the number of channels as CBers. This "frequency spectrum" includes frequencies that are normally local, and some that normally provide long-distance communications. For hams, communications range is limited only by the laws of physics. In fact, there's a trophy just waiting to be awarded to the first ham to communicate from Mars. It's just a matter of time.!

Hams use their radios in their vehicles, to talk with other travelers on interstate highways or to catch up with friends on the daily commute to work. Cellular phones have brought mobile telephone service to many, but Amateur Radio operators can also connect to the phone lines—and there's no usage fee.

Can Hams Do Anything They Want?

What *can't* hams do? For one thing, hams can't conduct business over the air. The amateur service is strictly noncommercial, and hams are careful about discussing anything related to business on the air. Hams also can't accept payment of any kind for operating their radios.

Who tells hams what they can and can't do? The Federal Communications Commission (FCC) makes the rules and regulations governing the amateur service in the United States. (The FCC also governs broadcasting—and many other radio services—in the US.) There are international regulations, as well. These regulations tell hams which frequency bands they can operate on, how much power they can use and which communication methods they can use. In addition, the FCC makes the rules governing testing and licensing in the US (more about that in a later chapter).

You can talk to people on a CB radio, but your range is normally limited to a few miles.

For hams, communications range is limited only by the laws of physics.

Serving the Public

In their pursuit of public service, Amateur Radio operators provide their own equipment, knowledge, time and effort at no cost. They ask for nothing in return but respect for their hobby and preservation of their allocations—the frequencies they're authorized to use. Ham radio is a noble endeavor, and one the world's hams are rightly proud of.

"HAM"—AN EXPRESSION OF HONOR

You may wonder why Amateur Radio operators are commonly called *hams*. No one really knows, but the tradition dates back many, many years.

It's likely that in the early days of wireless (just after the turn of the century) commercial telegraphers heard amateurs on the air from time to time. Perhaps they teased the amateurs by referring to them as "hams," the way one might needle an actor. Maybe they were poking fun at a few amateurs who sent sloppy-sounding code. (Code operators refer to a person's code-sending style as his "fist," and "ham" may have been a shortened form of "ham-fisted," or clumsy.)

Amateurs have no reason to be embarrassed about their forerunners' prowess. After all, most of the commercial and military "pros" cut their teeth on Amateur Radio. In addition, hams were responsible for virtually all the significant early developments in wireless communications. In fact, Guglielmo Marconi, the inventor of wireless communication, considered himself to be an amateur.

The precise origin of the term "ham" is destined to remain a mystery. Today, radio amateurs all over the world are justifiably proud of their heritage and are honored to call themselves hams.

2

What Do Hams Do, Anyway?

Once you get into ham radio, you'll find any number of activities to get involved with—on the air and off. Most hams aren't "just hams." They're DXers, contesters, VHFers or QRPers. They're satellite enthusiasts, traffic handlers or builders. A few pursue all these hobbies within the hobby. Most limit themselves to two or three at any one time. Hams are always finding new ways to enjoy the many facets of their hobby. You'll read about some of them in this chapter.

With all the ham activities to choose from, it won't be long before you're calling yourself a DXer or a QRPer, too! (You'll find out what these abbreviations mean later on.)

A GREAT EQUALIZER

Amateur Radio is open to anyone interested in pursuing it. Radio is a great equalizer. When you talk on

While getting away from it all on Canada's remote Manitoulin Island in Lake Huron, this ham set up a makeshift station and enjoyed his radio hobby. With the help of his two daughters (aged 7 and 9) and their friend, he built this antenna from cedar branches, hemp and snare wire.

the air, no one knows or cares about your age, color, religion, ethnic background, profession or appearance. Hams *are* interested in how your signal sounds (and how theirs sound), what kind of equipment you're using, what other activities you enjoy or how many countries you've worked (contacted).

People with disabilities use Amateur Radio to make friends around the world—without leaving home! Blindness is no barrier, as devices can translate text-based ham communications into the spoken word. Hearing-impaired hams use modes of communication that don't rely on sounds. Hams who are severely paralyzed can use devices that allow them to type or send Morse code by moving facial muscles or blowing puffs of breath to operate a radio. (See the Where to Find It section at the back of this book for contact information about the HANDI-HAMS, an organization that helps persons with disabilities get started in ham radio.)

DXING: MAKING FRIENDS AROUND THE WORLD

Ever thought about how nice it would be to travel around the world, learning about distant places by talking to people who live there? Who hasn't! Ham radio can be your ticket to worldwide adventure. That's just one of the reasons *DXing*—contacting other hams in faraway places—intrigues so many people.

You'll hear hams say things like, "I worked some DX last night. It was Rachel at 4Z4QP in Israel." Hams who love making such contacts are called *DXers*. Most of them can tell you precisely how many countries they've contacted. Some American hams consider any non-US ham to be DX. Others refer to stations outside North America as DX.

Hams aren't the only ones who "chase DX." Shortwave listeners (SWLs) tune in foreign music, news and other programs from around the world. Many SWLs get interested in Amateur Radio by listening. At some point, they get the urge to *talk* to people in other lands. Often, hams are part-time SWLs—they enjoy finding out what goes on between the different ham bands.

The DX Adventure

In the early days of wireless communications, it was a notable feat to conduct a two-way radio conversation

Ham radio attracts people because of the sense of adventure—you never know who you're going to meet next. Nothing matches the thrill of talking to a ham in another country for the first time.

with a station across an ocean. Today, the world is a much smaller place. You see live reports from halfway around the world on the evening news, and you can call almost anywhere in the world on the telephone.

What's missing from these other means of communication? The personal angle. Sure, you can *see* people from other places on TV, but you can't talk to them. You can't compare notes about your job and family, and you can't ask about their latest piece of equipment or new antenna.

Sure, telephone communication is two-way, but what attracts many hams to DXing is a sense of *adventure*. You never know who you're going to meet next. You might talk to a dentist from Berlin one day, and a Peace Corps worker in a remote village in the African country of Kenya the next. It's easy to get started; there are lots of DX hams to talk to. Nothing matches the thrill of talking with a ham in another country for the first time. You can even brag to friends about the "new one" you just bagged!

Part of the DX game is collecting contacts with as many different countries as you can. Many of these countries you've heard of—England, Japan, Hungary. Others are uninhabited spits of land like Bouvet Island in the South Atlantic and Rotuma Island in the Pacific.

QSLs: Postcards You'll Want to Save

Once you've worked a DX station, it's time to send—and with any luck, to receive—a *QSL card*. A QSL card is a postcard that serves as a record of the two-way contact. "To QSL" means to mail your QSL card to another ham to confirm a two-way radio contact.

It's a fine achievement to talk to 50 or 75 different countries, but how do you prove to your friends that you've done it? Simple: You show them the 50 or 75 QSL cards you've received from those you've contacted. The card includes information about the specific contact: both stations' call signs, the band or frequency used, the date

QSL cards serve as a confirmation of a two-way Amateur Radio contact—but they can be much more. This attractive card came about when two Amateur Radio clubs, one in Portland, Oregon and one in Sapporo, Japan became "sister clubs."

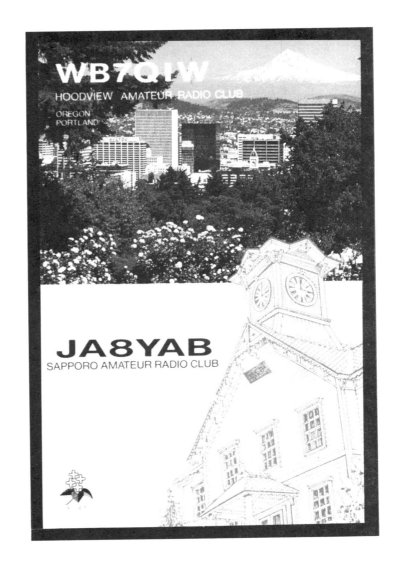

and time, and a signal report (how well you heard the other station).

Most hams have custom-made QSL cards that include additional information such as radio and antenna used, nicknames, other hobbies or interests, on-air achievements, or cultural and geographical notes about their locations. There may even be a handwritten personal note from the operator. It's little wonder that most hams collect QSL cards as mementos of their DX contacts.

We've been discussing high-frequency (HF) DXing, but DXing isn't limited to the HF bands—those ham bands below 30 MHz in frequency. There's DX-type activity on the normally local VHF, UHF and the micro-wave bands, too. Sometimes, VHF and UHF radio signals are affected by weather conditions such as cold fronts or by disturbances in the magnetic field that surrounds the Earth. When this happens, contacts of 1000 miles or more are possible on bands that are usually limited to line of sight.

CONTESTING: YOU CALL THIS *FUN?*

There are so many on-air operating contests that there's bound to be one that's just right for you!

Amateur Radio contesting is for those who love to compete—even if it's just competing with your own score from last year. The rules vary, but most contests have a common goal: to try to contact as many stations as possible, in as many different geographic locations as possible, during the contest period.

Sometimes you try to contact as many DX stations as possible. Other contests allow you to work anybody, anywhere. Some cover many HF or VHF bands; others are held on a single band.

"CQ contest!" fills the airwaves on contest weekends, as thousands of hams compete for certificates or other awards that attest to their operating skill and station efficiency. Sound a bit intense? Many contests—and the hams who take part in them—are, but there are beginners' contests, too. The ARRL Novice Roundup, for example, is only for those with Technician and Novice licenses.

Some hams take contests seriously: They will do little else during the weekend except operate. (Eating and drinking often take a back seat to nabbing that weak station in South Dakota or Albania.)

As these views of the same contest show, you can take any of several approaches to the art of contesting. Despite his relaxed position when the photo was taken, 6-year-old Tony (bottom) finished first in his category! Contests are a fun time that will net you several rewards, the first of which is the simple satisfaction of participating. Second, it's an excellent way to improve your operating skills. Third, it's a way to build your states-worked and countries-worked totals. DX contests offer the opportunity to work many DX stations in quick order.

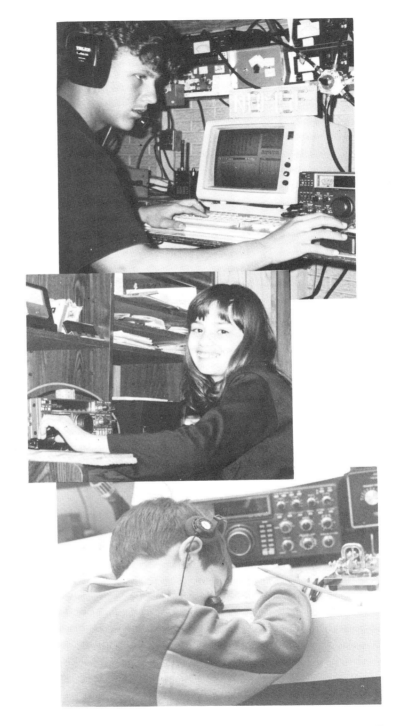

This gent brought a hard-working friend along when he operated from a mountaintop in Rocky Mountain National Park in Colorado.

VHF enthusiasts can join the fun by jumping into VHF and UHF contests. An exotic spot for a VHF/UHF contest station may be a high mountaintop in a location without regular VHF/UHF activity. Pack an antenna, radio, battery or power generator and other essentials into the family car and head for the hills. You'll soon find out how exciting it can be to have 10 or 15 stations calling you at the same time!

CHASING AWARDS AND CERTIFICATES

How many awards can you qualify for? No one knows, but "thousands" is a good guess. The award you're likely to obtain first requires only one contact! Start a conversation with someone on the air and talk for at least 30 minutes. Just hold a friendly conversation about anything, "chewing the rag," so to speak. You've just qualified for your first award, ARRL's Rag Chewers' Club certificate. That's all there is to it. That's not even the easiest certificate you can earn—the League also issues a First Contact Award.

This handsome award is sponsored by the West Siberia DX Club. You can earn it by contacting several stations in that part of the world—including up to three drifting stations in the Arctic Ocean!

Most awards require more effort than a half-hour chat. They also usually require you to have proof of contact in the form of QSL cards.

One of the most popular awards is ARRL's Worked All States (WAS). To earn it, you simply need to contact—and receive confirmation from—stations in all 50 states. Think that sounds awesome? Many hams have earned a WAS variation: 5-band WAS. To qualify for this one, you need to have confirmed two-way contacts from all 50 states on five different ham bands!

New states will mount up quickly when you first get on the air. The first 10, 15, 20, 30 or more are easy. For most amateurs, though, tracking down that last handful of states is a challenge. At one point, I had concluded that no one lives in New Mexico or Utah. A buddy said, "Nah, those are easy. South Carolina and Delaware are the hard ones." I finally figured out that whatever state you don't have confirmed yet is the "hard one."

The Worked All Zones (WAZ) award, sponsored by *CQ* Magazine, is for those who have confirmed contacts with 40 "zones" (as defined by *CQ*) of the world. Earning this award is a great way to increase your country total.

Some hams call themselves "county hunters." As the name implies, their goal is to work all 3076 US counties—no small feat, but it's been done by several hundred hams.

SPECIAL-EVENT STATIONS

When a ham club or other group decides to commemorate a historical or other public event, it may set up a *special-event station*. What kind of event? It may be the annual Buckwheat Festival in Kingwood, West Virginia, the 250th anniversary of the University of Pennsylvania, the reenactment of a famous Boer War battle or the 500th anniversary of Columbus' voyage to the New World. Contact the special-event station and you'll get a handsome QSL card or certificate. Hundreds of these operations are listed in the ham radio magazines each year.

There are awards for other activities besides making contacts. For example, the ARRL offers a certificate for proficiency in receiving Morse code. For this one, you don't even need an Amateur Radio license. You earn the basic award by demonstrating that you can copy one full minute of Morse code with no mistakes.

Those who contacted a special-event station set up to commemorate the 1990 World Economic Summit in Houston received this certificate.

WHY MORSE CODE?

Although you don't have to know Morse code to earn an Amateur Radio license, you'll probably want to at some point in your ham career.

Why would anyone want to learn such an outmoded and difficult means of communication? Two reasons come to mind immediately: It's not outmoded, and it's anything but difficult!

There are plenty of other good reasons as well. For starters, code gets through when voice modes don't. If you're fighting to be heard over noise or interference, there's nothing as effective as code.

Also, you'll need to learn the code if you want to earn a license that gives the HF privileges that hams use to communicate around the world. The code-free Technician license has no Morse code requirement, but all the other license classes do.

One more thing: Once you get into code, you'll probably surprise yourself and find you're enjoying it. Computer programs or cassette tapes make it easy to learn and practice at your own pace. It's like learning a new language, with a rhythm and a feel all its own.

If all that weren't enough to get you interested in Morse code, you can earn a code-proficiency certificate from ARRL for copying code sent over its station, W1AW. Several times a month, W1AW

Joseph Parskey, of Scituate, Massachusetts, has been an active CW operator for 27 of his 38 years. "I have always found Morse code to be one of my great loves," he says. Many hams start by using a straight key. After gaining some experience, you may want to use a computer to send code.

broadcasts special *code-proficiency runs.* Copy a minute's worth of code from the highest speed you can and submit your original paper to ARRL. They'll verify that you have the solid minute of copy, and a handsome certificate will be on its way to you.

The ARRL (see the Where to Find It section of this book) has more information on how to learn Morse code.

When there's an accident or disaster—like a chemical spill or an earthquake—radio operators trained in emergency communication flock to the scene. Carrying hand-held or other portable radios, they help local disaster and law-enforcement personnel by providing backup communications. When telephones and electrical service are knocked out, ham radio can always be on the air, thanks to battery power. These communications skills pay great dividends—in terms of helping to save lives and property, and in terms of goodwill.

Hams have the exclusive use of many valuable radio frequencies. A major reason for this is the goodwill they

The public service that hams provide can take many forms. This California ham, Bindy Beck-Meyer, shoots ham radio television pictures from a helicopter to assist the Highway Patrol, the Marin County Sheriff's Department and the Department of Forestry during emergencies.

When emergencies or natural disasters strike, ham volunteers don't wait to be asked—they take to the airwaves and provide communications support to local relief officials. Hundreds of hams assisted in the aftermath of the 1989 earthquake that struck the San Francisco Bay area.

generate by providing—at no cost—public service to their communities.

You've probably seen it on the evening news—hams often provide the first solid information from a disaster site. Hams hone emergency-communications skills through message handling or "passing traffic," as it's called. As a free service to the public, hams relay non-commercial messages across town, across the country or around the world. Although few of the messages are of real importance, traffic-handling serves two purposes: (1) It prepares hams for emergencies that might require their participation, and (2) it's an effective way for hams to remind the public that they're ready, willing and able to help when there's a real emergency.

Amateur Radio public service also shines during events such as marathons, dogsled races, charity walka-thons, parades and boat races. Every year, for example, more than 25,000 runners gather in New York City for the New York Marathon. Hundreds of hams volunteer to work side by side with race officials to make sure

Amateur Radio is a family affair for 9-year-old Dana. Her dad's a ham, too.

injured or ill participants and spectators receive medical attention.

TAKING AMATEUR RADIO WITH YOU

You *can* take it with you. In fact, most hams, do. Whether it's in your canoe, your pickup or your backpack, your radio can be your constant companion when you're away from home.

Bill Savarese, of Smithtown, New York, is a policeman. He and his wife Janice love to take their two children camping in upstate New York. Bill brings his 2-meter FM mobile rig and chats with other hams as he drives to their favorite campsite. He also takes along an HF transceiver that operates from his car battery, using a simple wire antenna. During the day, he strings the antenna through carefully selected tree branches. After Janice and the kids turn in for the night, Bill spends an hour or two on the air—making friends with other hams around the country and around the world.

Like to backpack? Unless you're an Olympic weightlifter, you'll want to choose the smallest and lightest equipment available. Then there's the other side of the coin, as exemplified by the Wally Bynam Radio Club. Members of this club are hams who own recreational vehicles (RVs)—some with high-power radios and pneumatically driven telescoping towers!

Amateur Radio is a useful tool for the modern family on the go. My wife, Sally, is a sales representative. Although she can't sell products with ham radio, she can put it to a more practical use. She works many evenings, and she's happy she has a mobile radio neatly installed in her company vehicle. If she has trouble, help is only a radio call away—to me or our teenage daughter. Ham radio makes it easy—and fun—for our family to stay in touch.

Extending Your Range with Repeaters

Bruce Lomasky, a computer consultant from West Hartford, Connecticut, uses his 222-MHz rig to ease the boredom of traveling from one client to another. He can

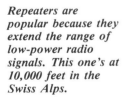

Repeaters are popular because they extend the range of low-power radio signals. This one's at 10,000 feet in the Swiss Alps.

usually find someone to talk to, though his "drive time" isn't during the regular commuter hours most of us are familiar with.

To extend his range while in his car, Bruce operates through a ham radio relay station called a *repeater*. Repeaters allow hams using low-power, mobile or hand-held radios to talk up to 50 or even 75 miles away. (The actual range depends on where the repeater is located.) During drive time, the VHF and UHF bands come alive across the country as amateurs use repeaters as they travel to and from work.

The VHF/UHF repeater has done more to draw local hams together than any other single development in Amateur Radio. Tune around the most popular VHF and UHF ham bands at 144-148 (2 meters), 222-225 and 420-450 MHz, and you're bound to find several local repeaters. You'll find folks across town talking about whatever interests them—their new antenna, a club picnic, a recent vacation or DXpedition, a TV special or the best recipe for chocolate-chip cookies.

Repeaters are ideal for finding your way. There's almost always someone listening for a call, so if you need directions, you can just get on the repeater and say so. Hams go out of their way to be helpful.

Mobile ham stations can also be lifesavers in time of need. Using a telephone-interconnect system called an *autopatch*, hams who come across traffic accidents, disabled vehicles and the like can summon police or rescue authorities for rapid response.

HF DXing, Boating and Flying High

Most mobile stations use the VHF and UHF bands, but many hams enjoy the thrill of talking to the world while driving, via the HF bands. Modern transceivers are small enough to fit easily in most vehicles. It's quite a conversation starter to arrive at the office and mention to your coworkers that you just finished chatting with a teacher in Mexico or a doctor in England!

Amateur Radio fits in nicely with other leisure-time pursuits. For the boater, ham radio is a window on the world—and a reliable backup emergency-communications system. Several boating magazines feature regular ham radio columns.

Not all mobile ham stations are this elaborate; all you really need is a ham transceiver and an antenna. The simplest way to operate from a vehicle is to use a hand-held radio like the one next to the steering wheel.

DXPEDITION: THE ULTIMATE HAM VACATION

There are places where stations and operators are as rare as hen's teeth—Clipperton Island, for example. Lying about a thousand miles off the southwest coast of Mexico, Clipperton is uninhabited. There are no homes, no factories, no electrical lines, no supermarkets, no airports. Like many other remote islands, Clipperton is covered with nothing much except plants, birds, reptiles and their "by-products."

Hams, though, have a special relationship with this remote island: it counts as a "DX country." (Ham radio countries can be very different from the usual definition of "country.") If nobody lives there, how can you work it? Periodically, groups of hardy (some say obsessed) radio amateurs plan a radio invasion of Clipperton (or any of dozens of other spots on the globe where there is little or no ham radio activity). Going off to an unusual location to operate a portable ham radio station is called going on a *DXpedition.* It's an exciting challenge to "activate" a radio-silent country.

A DXpedition to a rare spot is a major happening in

Not all DXpeditions are elaborate and expensive. This fellow, shown installing fish line in a tree, took part in a Poor Man's DXpedition to Martha's Vineyard, an island off the Massachusetts coast. The first-time DXpeditioners made 1000 contacts! (In case you're wondering, the line is used to pull a wire antenna into a tree.)

Amateur Radio circles. Plans are announced well in advance, and hams around the world track the expedition's progress. Tension mounts as the group approaches its destination. Will we be able to hear them? Will they hear us? What time of day do we have the best shot? Which band is best? These questions and a thousand others run through the mind of the DXer who needs (hasn't yet made a radio contact with) the rare one. The bands are alive

Activating a rare spot during a contest, as this gent found, can be a great way to combine the advantages of DXpeditioning and contesting. His operating skill and anti-bug supplies helped Jim win single-operator honors in the ARRL DX Contest from the tiny island of San Andreas.

Two American hams roughed it from this beachside location in the Caribbean nation of Belize. Somehow they found the motivation to make more than 3000 contacts.

with speculation and anticipation as the group approaches the desolate shores.

Have you ever watched a Presidential press conference on TV? You hear the roar of the reporters trying to get the President's attention, then a sudden silence as he picks one out of the bedlam and takes a question. As he finishes, the clamor bursts forth again until he recognizes the next questioner.

A horde of operators swarming over a DXpedition station's call sounds almost the same way—but worse. When the rare spot comes on the air, the airwaves come alive with a cacophony of call signs as thousands of hams try to contact the DXpedition. Voices bellow into microphones or Morse code "dits" and "dahs" beep from telegraph keys around the world.

The DXpeditioners sift through the chaos, searching for a recognizable call sign and invite that one lucky station to exchange some basic information, usually just call signs and signal reports. When the DX station indicates he's finished with the contact, the pandemonium resumes. Each complete contact may be as brief as 15 or 20 seconds!

Hams who can afford the costs involved—and who want to spend their vacation "on the air"—say there's nothing like it. Most, it's safe to say, would prefer to work the rare spot from the comforts of home.

Operating from a Cessna 182 Skylane at 130 miles an hour, a mile above Ocala, Florida is one way to attract attention. This flight instructor made 68 contacts on three different bands.

When this California ham went for a ride on the Goodyear blimp, he brought his 2-meter hand-held radio along. As it turned out, his operating time was limited. When the pilot learned the ham was a private pilot, he let him take the controls!

Private pilots find that it's easy to make long-distance line-of-sight contacts from 5000 or 10,000 feet in the air. Most ham/pilots use VHF/UHF bands, bypassing repeaters. (Although it's possible to operate through a repeater from the air, it's usually not a good idea. You'll be likely to activate—and tie up—several repeaters in addition to the one you're talking through.)

RAG CHEWING: GETTING TO KNOW YOU

"Rag chewing" means casual chit-chat. Many hams find that tuning around the bands with no particular goal in mind is a fun way to meet people. You might wind up speaking with someone on the other side of town, the country or the world!

In Amateur Radio, lasting friendships develop in spite of distance. Bill Frisch, of Lake Grove, New York, is retired. He spends much of his time talking with friends around the world. These friendships have blossomed to the point that he's had several European hams drop by for visits during trips to the US. When he took a vacation to Australia, his tour of the country was based on sights he wanted to see and old friends he wanted to visit—except this would be the first time to meet them face-to-face. Bill has little interest in contesting and the other goal-oriented aspects of ham radio; his thrill is in meeting new people and making friends on the airwaves.

According to the FCC rules governing Amateur Radio, one of its primary goals is to promote and enhance international goodwill. Although most hams around the world know enough English to have a brief conversation, even hams who speak different languages can get to know a little about each other using universally known ham jargon and abbreviations.

It's nearly impossible to describe your feelings as you speak with an exuberant ham in Romania or the former East Germany, hours after his country has been liberated from a repressive government. *Glasnost* is not an abstraction when you're able to talk directly with Russian hams about their society's formidable challenges and promising reforms.

QRP: THE ART OF FLEA POWER

QRP means operating with low power, and QRPers pride themselves on making contacts with as little transmitter power as possible. Compare a light bulb at 75 watts with the transmitter power some hams use to communicate around the world—5 watts, or even less!

Low-power operation is a challenge, and may not be the best choice for those getting started. Yet it's a rewarding and significant accomplishment to converse with amateurs in exotic, distant lands with the same output power as a home garage-door opener.

This battery-powered QRP station is ideal for operating in the woods. The whole thing fits in a small knapsack (in the background).

HAM RADIO IN SPACE

Reaching out, stretching beyond boundaries, "pushing the envelope"—however you describe it, it's what many hams strive for. The ham radio space program is a prime example of what can be achieved with a dream and a great deal of hard work.

Hams have their own orbiting satellites—and several space shuttle missions have had hams aboard who provided the contact of a lifetime as the shuttle hurtled overhead at 17,000 miles an hour.

The first public proposal for hams to have their own satellites came from Don Stoner, of Mercer Island,

Washington, in his *CQ* magazine column. It was published less than two years after the Soviet *Sputnik* had ushered in the space age. It was only two short years after that, in December 1961, that hams had their very own satellite—OSCAR I (for *O*rbiting *S*atellite *C*arrying *A*mateur *R*adio).

The first nongovernment satellite ever launched, OSCAR I was put together by resourceful California hams for a total cash outlay of less than $30. Its ejection mechanism was powered by a $1.15 spring from Sears! Nearly 600 amateurs in 28 countries reported hearing OSCAR I's low-power beacon transmitter, which sent HI in Morse code at a speed that varied with the spacecraft's temperature. After three weeks, the batteries ran down and OSCAR I went silent.

Since 1961, hams in several different countries have designed, built and launched more than 20 orbiting satellites. Today's OSCARs are sophisticated communications satellites that offer worldwide communications with simple VHF/UHF equipment.

You're probably wondering how a bunch of regular people can have their own private satellites. No, hams

*Those who contacted the Soviet space station **Mir** had a thrill of a lifetime—and this QSL card to remember it by.*

Soviet space station MIR

don't have a secret fleet of launch vehicles like a James Bond movie. They simply do what hams do best—scrounge parts, pour on the elbow grease, cajole assistance from commercial firms to supply specialized testing systems and make do with whatever works. Amateur satellites "hitchhike" rides on launch vehicles with government or commercial payloads. (If hams had to launch their own satellites, the cost would run into the millions of dollars.)

Shuttling Ham Radio into Orbit

There's more to ham radio space communications than the relay stations floating through the heavens. On a warm Florida morning in late November 1983, the space shuttle *Columbia*'s engines roared to life, blasting mission STS-9 into orbit. The power of *Columbia*'s engines rocked ground observers in waves of what sounded like a thousand cannons firing at once.

That shock wave was nothing compared to the one astronaut Dr Owen Garriott, who holds the call sign W5LFL, created in the ham community over the next two weeks. With NASA's permission, Garriott carried a portable 2-meter FM transceiver and window-mounted antenna aboard *Columbia*. In his off-duty hours, he made direct, personal contacts with hams all over the globe in the first ham operation from space.

Since STS-9, thousands of hams worldwide and schoolchildren attending live demonstrations in their classrooms have had the thrill of contacting a ham aboard the shuttle. Russia has a permanent ham station on the *Mir* space station. Hundreds of hams have contacted *Mir* and earned a prized QSL card as a memento. Can you imagine holding a radio the size of a cigarette pack in your hand as you sit in your living room talking to an orbiting astronaut?

Can you imagine holding a radio the size of a cigarette pack in your hand as you sit in your living room talking to an orbiting astronaut?

Let's make it clear up front: No law says hams have to tinker with their equipment. You can pull the plastic out of your wallet and buy everything you'll ever need for your ham station off the shelf. Many people do it that way. Then there's the tinkerer.

Charlie Helmick, of Parkersburg, West Virginia, has been a tinkerer from Day One. He has a complete HF ham station, but you'd seldom catch him operating it. What he likes best is to build things. He's curious about how things work. When his wife bought him a top-of-the-line programmable scanner a couple of Christmases ago, Charlie took it apart before he even turned it on! (Unlike the way small children often dismember their presents and reduce them to piles of wreckage, Charlie put his scanner back together and it works just fine.) His

A code-practice oscillator like this one makes an ideal project for a ham radio class. It's fun, and it provides a useful piece of gear for learning and practicing the Morse code. Before commercial ham gear became common, hams had to build their entire stations. Some still do.

This group, members of the Los Angeles fire department, built an antenna while they learned about ham radio in a Novice class. Their instructor (third from left) looks pleased.

Many hams find it satisfying to build an accessory, or even a complete ham station. If you can figure out which is the business end of a soldering iron, you can put together components of an Amateur Radio station.

biggest thrill is to put something together and fiddle with it until it works perfectly. When he's satisfied with it, he moves on to something else.

Another element of experimenting is designing and building equipment. Many hams find it satisfying to build an accessory, or even a complete ham station. If you can figure out which is the business end of a soldering iron, you can put together components of an Amateur Radio station.

The Resources section of this book lists some companies that offer kits with all parts and instructions. Those who build a kit find that they learn a great deal about how it works—knowledge that will come in handy in case of problems later on.

Ham antennas are an ideal way to try tinkering. Hams have tried countless arrangements of wire, aluminum tubing and mechanical frameworks to squirt radio signals into the ether. You can design your own, modify one that you buy, or build one from a description in one of the ham magazines or reference books. Any way you go about it, it's a source of pride to be able to tell your friends that you built some of the equipment in your station.

When she was 9 years old, our daughter Anita decided she wanted to get her ham license. She had watched her parents talking on the radio and decided she wanted in on it. (Kids as young as 5 have earned their Amateur Radio licenses, but it's still quite a triumph for a 9 year old.) When she passed her exam, I ordered a Morse code keyboard kit for her to build. We had a lot of fun as she put it together. She benefited from that experience both as a ham and as a student—it also helped her in some of her 8th grade classes.

COLLECTING HAM TREASURES

You name it, and somebody collects it—old bottles and record albums, figurines, plates with handpainted celebrity pictures on them, beer cans, refrigerator magnets, antique furniture and paintings. Ham radio also has its share of collectibles. The types of things people collect and the reason they collect them varies widely.

Most hams have a QSL card collection. Walk into some ham shacks and you will find the walls covered with cards from hams all over the country and the rest of the world. Head across town to another ham's station, and you won't see a single card on the wall—but you'll find a stack of shoeboxes full of them in the corner. Inside, you may discover cards from all over the world—the Falkland Islands, Kuwait, Botswana, China, American Samoa, Vanuatu, Reunion Island and Antarctica. The cards hold warm memories of pleasant chats with people halfway around the world. Who wouldn't treasure them?

Some hams collect old equipment and restore it. Walk into their shacks and you're transported back in time 50 years, to the days when radio was king. Behind the polished oak cabinets, tubes emit a warm glow that casts eerie shadows on the wall. There's a thrill to bringing an old piece of gear back from the junkpile, and

Some hams collect old equipment and restore it. Walk into their shacks and you're transported back in time 50 years, to the days when radio was king.

This Ludlow, Massachusetts, radio amateur's impressive collection includes ham radio license plates from every state and Canadian province—and the first ham plate ever issued in the island of Aruba. Hams collect just about anything that interests them, from postage stamps with a ham radio theme to "wireless" magazines dating from the early part of the century.

What many folks don't know is that hams were experimenting with methods of transmitting moving pictures in the late 1920s and early '30s, well before the earliest commercial broadcasts.

thousands of hams partake of that joy. There are firms that specialize in providing hard-to-find components needed to restore vintage equipment. For some, trading equipment can be as much fun as restoring it.

Arnold Chase, of West Hartford, Connecticut, has an impressive collection of early TV sets. What many folks don't know is that hams were experimenting with methods of transmitting moving pictures in the late 1920s and early '30s, well before the earliest commercial broadcasts. Some of Chase's most prized trophies are one-of-a-kind homemade sets.

Simple things often become collectibles. Telegraph keys, some dating from the 19th century, are fast-moving items at ham flea markets.

Magazines, particularly old ones, are another target of collectors. The ARRL began publishing *QST* in 1914. *Radio,* the forerunner of *CQ,* came on the scene in the 1930s. Complete collections of the early editions of these magazines can be quite valuable. They can also be a source of entertainment. If you hunt diligently, perhaps you'll be fortunate enough to locate some rare old publications.

If you can get your hands on one, curl up with a vintage copy of *Amateur Radio News*, *Radio World* or *Radio News* for a trip back in time. Read the tidbits and the articles. You'll soon find yourself caught up in the romance of radio in the early days.

IS THIS AMATEUR RADIO?

Ham radio is everything we've discussed in this chapter—and more. Even today, when it's possible to pick up a telephone and dial most parts of the world directly, chasing DX on ham radio has lost little of its glamour.

Yet Amateur Radio is more than this. It's talking with your buddies as you drive to work in the morning. It's checking with your spouse on the way home to see if going out for dinner would be a better idea than cooking. It's the unequaled thrill of chatting directly with an astronaut as the space shuttle streaks through the sky. It's volunteering to provide communications help at a charity event. It's arranging for supplies to be shipped to a starving family in the aftermath of a killer hurricane.

Ham radio is more than you can read in a book. It's a service that carries on 75-year-old traditions while staying aligned with the razor edge of modern communications technology. Hams can be found in almost every country of the world.

It's a hobby that excites inner-city schoolchildren, millionaire celebrities, homemakers, office workers, royalty and entire families. It transports those unable to travel themselves to places they can only dream of visiting in person. It brings together crusty old-timers using home-made vacuum-tube radios from yesteryear and teenagers using today's compact, digital hand-held transceivers connected to lightning-fast microcomputers. It links American college students, African Bushmen, Polynesian fishermen, Japanese engineers, European factory workers, South American mineworkers and Russian farmers in a warm family of close-knit brothers and sisters.

Ham radio has to be experienced to be appreciated.

Ham radio is more than you can read in a book. It's a service that carries on 75-year-old traditions while staying aligned with the razor edge of modern communications technology.

3

Getting Together: Ham Radio as Social Lubricant

Whether they're flying the Goodyear blimp, planning a DXpedition to Rotuma Island or improving their stations in time for the next contest, hams are an active bunch. As a ham, you can work the world from your armchair—but few hams are content to stay there for long. In this chapter, we'll look at some of the ways hams get to know each other—face to face.

TAKE A CLASS, THEN JOIN A CLUB

Most people who want to earn their first ham license find it's easier to do in a class than it is alone. Classes are usually sponsored by ham radio clubs. Club members serve as volunteer instructors. In a class, you'll get to meet several club members and perhaps a guest speaker or two. You'll find out what ham radio is all about in your local area. Best of all, you'll be likely to make some lasting friendships.

After you have your license—or even before—you'll probably want to join a local club. It's an ideal way to meet people with similar interests, people who understand the joys of Amateur Radio and enjoy talking about them. You'll get to know people to turn to if you have questions about setting up your station. And you'll get to pick someone's brain before you spend money on a new piece of equipment.

Ham clubs offer a superb way to fit into a new community if you happen to move. Hams are comfortable talking to people from all over, and they share common bonds—you'll be welcomed as part of the group.

Many hams make some of their closest lifelong friends through their clubs. Jennifer Roe belongs to several radio clubs in the Santa Barbara, California, area. She says, ''We hams have to stick together. Clubs are often our 'extended family' in times of crisis or need. They become good friends, and you really see that when a member has a family tragedy or needs a helping hand. Ham club members have helped me more than any other group. And we get involved with a lot more than ham-oriented activities. Clubs are the heart and soul of Amateur Radio.''

This happy group of Texas hams, members of a net that meets on the air once a week, also get together each year in person. "We don't feel like strangers," one of them commented. "We feel like sisters—like family. And we're interested in what's happening to each other. I think that's what holds us together."

Being an active member of a club means you may socialize with auto mechanics, corporation presidents, high school students, engineers, sales people, retirees, mail carriers, police officers, teachers and firefighters. You never know who you'll meet at a club meeting. Joe Walsh was a ham long before his days with the Eagles rock band. He once hired a helicopter to transport a blind ham to an island mountaintop—where they both helped their club, the Santa Barbara Amateur Radio Club, to a 5th place finish in a UHF contest. Lt Colonel Ken Cameron, who has operated ham radio from space, teaches licensing classes to other shuttle astronauts at the Johnson Space Center Amateur Radio Club in Houston.

The American Radio Relay League (ARRL) will get you in touch with a local Amateur Radio club. (See Where to Find It at the back of this book.) There are about 2000 clubs affiliated with ARRL, and many others besides. Odds are, you'll find one nearby that's right for you.

Clubs Come in All Flavors

Clubs are like people; every one is different. You may prefer club meetings that feature a guest speaker or video/slide program, with a minimum of "business." Try visiting several clubs before deciding which to join.

General-Interest Clubs

Some clubs apply their energy to diverse activities. Their motto might be "variety is the spice of ham life." Most meet once a month. There will likely be a short business session, followed by a presentation or guest speaker. One month, there may be a discussion of airborne search-and-rescue operations. Another program may be on wire antennas or how to get the longest life out of batteries for hand-held transceivers. The next month's program may be a video on the latest exploits of hams on the space shuttle. The following month could be a slide show by a ham who went to a chalet in the Alps to operate in a contest. Another program might feature a speaker from the National Weather Service, showing the group how to monitor and send in reports of severe weather.

Most clubs publish newsletters to keep members up to date on club affairs. A well-written and -produced newsletter can be a centerpiece for a club. Helping with the club newsletter is a great way to become active.

Another place for club members to meet is on the air. If the club sponsors a local repeater, it will probably be used for this on-the-air meeting, or *net,* as it's called.

General-interest clubs may offer other activities, too.

Licensing classes: Many clubs offer a course leading to a beginning ham license (or a higher-class license). A licensing class is a great way to master the material you need to pass the exam and to learn top-notch operating techniques. You'll also meet other people just starting out. Once you're licensed and have some on-the-air experience, you may want to help teach a class.

Promoting ham radio: Successful clubs let the community know they are there. It's good public relations

Many ham radio classes give students a chance to build something. This youngster is putting the finishing touches on a code-practice oscillator, used to practice sending Morse code. Its cost: $5!

TEN-TEN: A CLUB BASED ON FREQUENCY

A nationwide group popular among newcomers to ham radio is the Ten-Ten International Net Inc. It was formed in southern California to foster interest in and activity on the 10-meter ham band. Local chapters sprang up across the US. As 10 meters has grown in popularity, the organization has thrived.

Almost 100,000 hams from around the world have joined Ten-Ten, whose sole purpose is to have a good time operating on the 10-meter band. When you get your Technician Plus or Novice license, you'll be able to join, too. The Resources section at the end of this book has more information about Ten-Ten.

and it helps ferret out new hams. Many clubs find that setting up a station and operating it in the local shopping mall is an ideal way of bringing the public into direct contact with ham radio.

Special-event stations: Similarly, the public gets to learn a little about ham radio when a ham club joins with other community groups to celebrate a local event. Each year, the Holland (Michigan) Amateur Radio Club operates a special-event station during Tulip Time. Hams contacting the Holland ARC station receive an attractive commemorative certificate.

Scholarships: Clubs that can raise sufficient funds often sponsor scholarships for worthy local students. It's another effective way for a club to support its local community.

Special-Purpose Clubs

Some clubs focus on one area that its members are most interested in.

Public-Service Clubs

Whether it's a March of Dimes walk-a-thon or your town's Memorial Day parade, there are many opportunities for clubs to help with local (noncommercial) communications. If the event is big enough, like the New York City Marathon, several clubs will join forces to provide the communications.

Repeater Clubs

Some club activities—like sponsoring and maintaining a repeater—last all year. Clubs that sponsor repeaters usually form a committee to oversee its operation. (If you use a repeater regularly, most clubs expect you to join. It's the right thing to do.) Some general-interest clubs sponsor repeaters as well.

Packet Radio Clubs

The tremendous growth of computer-to-computer packet radio has spawned many clubs and organizations.

Many clubs hold annual dinners as a means of getting spouses and children involved in a club activity. When the Alabama DX Club held its annual banquet, it took the opportunity to make a visiting Russian ham (left) an honorary member.

Some are informal groups that may get together for pizza once a week, while others are more formal.

The Packeteers of Long Island meets each month. Its goal is to improve computerized packet radio operation on Long Island at all levels. Although the group coordinates sophisticated packet networks, membership is open to anyone who has an interest in packet radio regardless of experience.

DX Clubs

The DX club provides DXers with a forum for exchanging DX information, tips and "war stories." The Long Island (New York) DX Association (LIDXA) meets on the third Friday evening of each month at a local college. After a short business meeting, there's an open discussion of upcoming DX events. Finally, there's a program on a DX-related subject.

The LIDXA owns and operates a 2-meter repeater used for DX spotting. Members monitor the repeater while they're working DX stations on the HF bands. When a member finds and works a DX station, he or she calls the other members on the local repeater to let them know the call sign and frequency of the DX station. Once a week, the club holds a DX net on the air. The net is a structured affair that provides members with detailed information on current and anticipated DX operations.

In contrast, the Central Ohio DX Association meets every three or four months for dinner in the Columbus area. Profit from the meal is used to offset the cost of mailing meeting notices out to the members. After dinner, there's a short DX program. The club has no major expenses (such as a repeater), so there's no need for dues.

Contest Clubs

Contest clubs encourage their members to participate in all major contests and as many minor ones as they have time for. After the contest, most contest sponsors tally

DAYTON: HAMFEST HEAVEN

The Dayton HamVention is just a hamfest—only more so! Each April, about 30,000 hams descend on Dayton, Ohio to catch up on the latest technological breakthroughs and news of old ham friends (many of whom meet at the HamVention every year), and generally enjoy being immersed in ham radio for an entire weekend.

This giant exhibition area is only a small part of the largest ham get-together in North America—the Dayton, Ohio HamVention, known to hams around the world simply as Dayton. This monster hamfest is a great place to meet ham friends old and new.

There's a mystique about Dayton that has built its reputation to legendary proportion. The Dayton Amateur Radio Association (DARA), which organizes the HamVention, may be the most famous radio club in the world. "Dayton," as it's known to hams around the world, is the largest hamfest in North America and second largest in the world (Japan holds a larger one). Manufacturers schedule spectacular new product releases, and hams from all 50 states and most foreign countries schedule vacations to coincide with *the* ham radio event of the year.

Those who have been there tell new hams that "You've gotta go to Dayton. Every ham has to go. You can't imagine it until you've been there!" Many hams go every year. Ask a group of Dayton veterans why, and you probably won't get the same answer from any two.

From the enthusiasm of those who have been there, you have to wonder if there's something almost supernatural about it. And there is. It's an overwhelming, exuberant affair that exhibits the chaotic frenzy of Mardi Gras—and it's all about Amateur Radio.

Sooner or later, almost every ham *does* go to Dayton. Why? The Dayton HamVention is the ultimate hamfest. It's to ham radio what Times Square is to New Year's Eve—every ham should try to go at least once just to experience it.

the scores for all the club members so one club can compare its score with others.

Hams who join contest clubs are competitive. Keeping a contest log of stations contacted, on a personal computer or on paper, is an excellent way to help the club's contest effort, and to learn contesting.

Hamfests

Aside from membership dues, where does the money come from for repeaters and other club activities? An enjoyable way for a club to earn money to support a repeater or other club project is to hold a *hamfest*—a gathering of ham radio operators that's part flea market and part convention.

The club hamfest committee arranges a suitable location (such as a meeting hall, drive-in movie theater, school gymnasium or conference center), schedules a date (normally a Saturday or Sunday), invites local and regional businesses that sell the things hams like to buy, and prays for good weather.

On the day of the hamfest, club members show up early to make sure everything is set up correctly. Coffee is brewed, refreshments are prepared and members are deployed to patrol parking areas, assist vendors, man the doors and admission tables, and tie up loose ends.

When the doors first open, it's like being on the floor of the New York Stock Exchange. Money changes hands at a frenzied pace. You can spot the serious buyers—they come prepared with backpacks and luggage "wheelies" for carrying away their treasures.

There's more to a hamfest than just buying and selling. It's a chance for friends to get together and have an "eyeball QSO" (face-to-face chat). Hamfests draw more people than club meetings, and from a wider area, so they're a great chance to meet old friends and make new ones.

Large hamfests have talks and seminars on various aspects of the hobby. They are ideal ways to learn more about ham radio.

Picnics and Parties

When its members tire of committee reports and other formal business, clubs organize events that bring spouses and offspring together. Some clubs have several get-togethers during the year. Ham radio is a family activity, and club picnics and holiday dinners are a great way to get that message out.

The Suffolk County (New York) Radio Club has found it worthwhile to suspend meetings during July and August. In place of the August meeting, the group meets for a family picnic, inviting its members' families—and members and families of neighboring clubs. Picnic food abounds. Children chase frisbees across the field.

In December, the group holds a holiday dinner. In recent years, the catered buffet-style dinner has won out over going to a formal sit-down meal at a restaurant. It's

A potluck supper is a great way to get club members and families together, as the Augusta (Maine) Amateur Radio Club found out.

a time to bring family and friends to the club for an evening of fun, friendship and good cheer.

FIELD DAY: RADIO FUN IN THE WOODS

On the last full weekend of each June, hams across North America set up portable stations away from sources of electrical power. Field Day, as it's called, is the operating event thousands of hams look forward to each year. Although the purpose is serious—to practice communicating under conditions that might be found after a major disaster—Field Day usually turns into a social activity that celebrates the joy of Amateur Radio—and ham ingenuity.

Sponsored by the ARRL (beginning in 1933), Field Day is a contest-style competition based on the number of contacts your group can make during a 24-hour period. The League publishes the results in *QST*, so everyone makes an effort to do well. Since the rules encourage friendly competition, contesters and noncontesters alike get caught up in the excitement of Field Day. It's a perfect mixture of camping, picnicking, socializing, honing skills for emergencies and getting on the air.

Although individuals operate Field Day, it's more fun as a club activity. Because the rules favor those who use portable equipment (that which isn't connected to permanent antennas or regular commercial electrical power), most groups operate from camp-type settings. On Saturday morning, there's a flurry of activity as people rush about setting up towers, tents, tables and radios. By noon, members have assembled most of the station.

These hams—father and son—find that Field Day is an ideal way to spend a June weekend together.

Field Day groups can be large, like this one, or individual—but most enjoy the teamwork found in club groups.

The point of Field Day is to set up a working station away from power mains. Many groups use gas-powered generators. Others are more creative. Left: a Michigan group used this antique steam tractor to power their station. Right: Field Day can be healthy, too.

This Field Day operation stopped at nothing to have an efficient satellite antenna.

Julie and Marsha operate Field Day with the Arkansas-Tennessee Radio Association.

These Field Day operators from Little Rock, Arkansas were so intent on making that contact they didn't even notice the photographer.

Resourceful club members have analyzed and solved quirky problems—"Which tree should we use for the south end of the 80-meter dipole?" and "Who was supposed to bring the generator, anyway?"

Getting the tower up in style, members of two Colorado clubs demonstrate that Field Day should be fun as well as an exercise in emergency-preparedness.

Not all Field Day sites are literally located in a field, but this one, in Quebec, sure is.

Many who take part in Field Day enjoy the camaraderie more than the actual operating. Between organizing, setting up and tearing it all down, Field Day provides an ideal chance to meet new people and get together with old friends.

With most of the gear unpacked, the Field Day fun is about to begin.

Field Day operators appreciate the help of an assistant—someone to log contacts and to spot dupes (stations already worked). You don't need a ham radio license to log, of course—but it helps to be familiar with call-sign format. Field Day also provides the opportunity to assist with set-up and tear-down. The more people who help pack up the gear Sunday afternoon, the less time before the group can head on its way and deal with going back to work the next morning!

4

Gadgetry and Technology: The Cutting Edge

Most of us are fascinated by technology. Some hams are satisfied to buy a good piece of equipment, put it on the air and use it until the silicon wafers in the transistors revert to sand. But most of us want more. There's always something new and different to try. "This new radio has 'bells and whistles' that weren't available a couple of years ago," they'll say. "I just *have* to have one!"

When my wife and I bought our first 2-meter (144-148 MHz) portable transceivers, they were state of the art—1973 style. About the size of a CB radio and weighing in at about 2 pounds, they featured rechargeable batteries and three pairs of crystals, with sockets for another three. With our complement of crystals (adding significantly to the cost), we could operate on all of six different repeater frequencies.

As long as we stayed in our local area, everything was fine. Traveling was another story—when the local repeaters used frequencies we didn't have crystals for, as they usually did, we couldn't talk to anyone!

Fortunately for gadget lovers, electronics technology has surpassed the hopes and dreams of the most wide-eyed optimists. Breakthroughs became commonplace during the rest of the '70s, accelerated into the '80s and

The Drake TR-22 2-meter transceiver was state of the art, 1973-style. Today's microprocessor-controlled 2-meter gear is far more sophisticated—and in many cases far smaller and lighter.

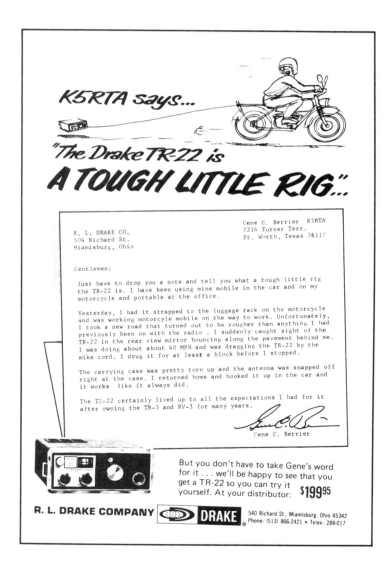

have shown no signs of letting up in the '90s. Today, the average portable 2-meter FM transceiver covers all 800 channels, has up to a few dozen memories, a memory "call channel" that can be tuned in with a single touch of a button—and more. Much more.

You can select from several ways of scanning across the band, across part of the band or through preselected memories. The receiver will even let you listen in on radio services on either side of the 2-meter amateur band, such as police, fire, civil-preparedness and aircraft. The radio has a fancy liquid-crystal display (LCD) readout like the ones on digital watches and calculators. Indicators tell you the status of many of the special features, in addition to the operating frequency. And this is all in a radio no larger than a pack of cigarettes!

It may be hard to believe, but this type of 2-meter FM rig isn't even hot stuff anymore. Several hand-held radios on the market today cover two completely separate frequency bands, such as 144-148 (2 meters) and 430-450 MHz (70 cm), with all these features included for each band. Modern hand-held transceivers can act as band-to-band repeaters, receiving on the 144-MHz band and transmitting on the 430-MHz band.

Gadgeteers shouldn't be surprised by any of this. If you think about the typical car radio, it wasn't too long ago that you had to wait for the tubes to warm up before you heard anything. Then car radios went transistor—and were suddenly cooler, smaller and sturdier, and came on instantly. Before too long, you had the option of FM as well as AM. Next came FM stereo, and then tape players. (What self-respecting gadgeteer would drive a car these days without a tape player?) Now, multi-changer CD players are becoming commonplace, as are high-quality speaker systems.

What's Next?

It may be hard to imagine what the research and development people have in store for us, but it's safe to say that technology will leap ahead ever more rapidly.

Some hams are so eager to harness electronics technology that they go looking for—and often find—undocumented features in their radios. Hams have found that modern transceivers can be made to work in ways not mentioned in the owner's manuals.

Anything not documented in the manual is probably of little use to most hams. But utility isn't the issue. This is a 1990s version of the spirit that drove hams to experiment with different tubes, frequency bands and antennas back in the 1920s and '30s. It's the same need to explore, to understand, to push a piece of equipment to its limit.

Imagine this curiosity carried up the scale an order of magnitude or two. Only recently considered esoteric, and put to use by only a few experimenters, *digital communication* is quickly becoming a commonplace ham activity. It's evolved pretty much the same way as the car radio and the portable VHF FM ham rig.

Hams are driven by the need to explore, to understand and to push a piece of equipment to its limit.

DIGITAL COMMUNICATION: ONLY THE BEGINNING ——

Communication is the process of passing information, or *data,* from one person or device to another. Talking is communication. Operating an automatic garage-door opener is communication. Watching television is communication. Reading this book is communication between author and reader. When you transmit on a radio, you're communicating a continuous stream of information to someone else. Voice transmission (also called *radiotelephone*) is a form of *analog* communication.

Analog means that something is a smooth, continuous copy of something else. The method of operation is *analogous*, or equivalent. For example, time lapses in a smooth, uninterrupted flow. Analog watches have hands that display the time in a continuous, circular sweep.

A brief discussion of how a voice signal gets from your station to others around the world might be useful here. The sounds you make are picked up by a microphone, which changes the sounds into a small current of

Figure 1—Clocks can be analog or digital. The one on the left is analog—it uses hands traveling in a circle to represent the time. (The 12 is comparable—or analogous—to midnight or noon, for example.) The digital clock on the right shows the actual numbers of hours, minutes and seconds. Amateur Radio operators use both analog and digital modes to communicate.

electricity that varies in relation to the pitch and volume of your voice. An infinite number of possible electrical waves are formed when the sound of your voice reaches a microphone, because every tone, inflection and part of a syllable causes the microphone to create a certain electrical pattern.

A radio transmitter picks up the weak electrical signal from the microphone, amplifies it to make it stronger and combines it with a powerful electrical current in the range of radio frequencies. This current passes out to the antenna and is radiated into the atmosphere. If it's a strong enough signal, a distant receiving set can detect the signals and reverse the process, creating a reproduction of your voice in a loudspeaker or earphones. Now your communication is available to the person listening to the receiver.

In *digital* communications, rather than sending a smoothly changing flow of radio waves from one station to another, a continuously changing stream of data is broken into small pieces that are encoded as a series of

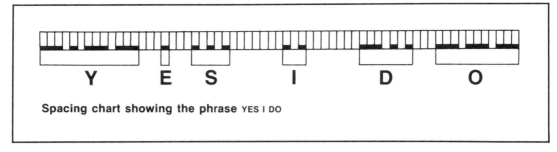

Spacing chart showing the phrase YES I DO

Figure 2—Ham radio operators use Morse code for many reasons, but two of the most important are that it gets through when other communications modes won't, and it's a unique skill that sets hams apart from other users of the frequency spectrum.

numbers. A surprisingly large amount of information can be encoded just by using two numbers. You know what the amazing power of a computer can accomplish, and it's all done by the manipulation of two numbers. The most common form of digital-information encoding uses just the numbers 0 and 1. It's called a *binary* system, because binary means something that has two parts.

A Bit About Morse Code

Morse code is a type of binary digital communications (a binary digit is called a *bit*). Hams using code translate words into a series of two signals: A long tone (dah) and a short one (dit). Using a switch or key to send these dits and dahs is called *telegraphy* (from the Greek roots *tele* for distant and *graphos*, or writing—writing over a long distance). Sending Morse code over the radio airwaves is called *radiotelegraphy*. An Amateur Radio transceiver generates a continuous wave of radio energy that's broken into dits and dahs for transmission. (The continuous wave used for Morse code gives us the common abbreviation for Morse code: *CW*.) The receiving station translates the digitally encoded characters (dits and dahs) back into the original text.

Originally, the encoding and decoding medium for Morse code was the human brain. Today, electronic devices can do this fairly well. You can hook up a device between a radio and a data communications terminal that lets you send Morse code by typing on a keyboard and receive code typed on paper or displayed on a monitor screen.

Digital and analog communications each have advantages and disadvantages. In a simple digital transmission, noise, static or interference can cause the receiver to miss a bit every now and then. Most of the time, however, the string of characters received correctly is still useful because the brain knows how to "fill in the blanks."

If you were to receive **MARY WILL BE COMING HOME TOMORRW AT 11 AM**, you'd know that the sixth word was supposed to be "tomorrow." On the other hand, you could miss a different bit that would make the message much harder to understand. If you received **MARY WILL BE COMING HOME TOMORROW AT 11 M**, you wouldn't be able to tell if the time was in the morning or evening.

Some data-transmission systems are less prone to such defects because they use a means of error detection. One way to do this is to add the numeric value of all the bits in a message and calculate the sum. This sum is sent at the beginning or end of the message, and the receiver compares the sum to the message received. If they don't match, an error must be present. If the message is being sent and received in "real time," the recipient could stop the transmitter and ask him to send the questionable segment of the message again. Just knowing that there's an error, however, doesn't always help identify what the error might be. If you received **MARY WILL BE COMING HOME AT 1 AM**, you would have no way to know the sender meant 11 AM.

In analog communications, these problems are often harder to identify. There's no such way to calculate a sum on a spoken transmission. A word count can ensure that the correct number of words have been received, but can't determine whether you made a mistake in copying

"Mary" for "Jerry." The brain steps in and fixes the problem if you happen to be talking about someone you know named Mary, but in almost any other case there could be a confusing result.

Imagine the possible consequences if a doctor advised an emergency medical technician to "give victim 15 units of morphine" and the message was received as "give victim 50 units of morphine."

Let Your Tones Do the Talking

Amateurs use other forms of digital radio communications aside from Morse code. If you understand how these work, you'll see the enormous potential they offer.

In the 19th century, someone developed a typewriter-like device that would send and receive electrical signals over wires to and from an identical device on the other end. The receiving machine decoded the signals and typed words on paper. This was known as *teletype* (*tele* means distant). Teletype is abbreviated *TTY*. In TTY, signals are sent as binary code (*mark* and *space*, equivalent to the computer's 1 and 0). Any two electrical signals can actuate the receiving terminal's output device. It could be two different tones, voltages or current levels, on/off pulses or other means.

In RTTY (the radio version of TTY) and another digital mode called *AMTOR*, two audio tones are used to represent digital 1s and 0s. A low tone may mean 0 and a higher-pitched tone may mean 1. A tone at one pitch represents the marks and another tone represents the spaces. Each letter of the alphabet, numeral, punctuation mark and special signal has its equivalent mark and space value. The sound that stands for **A** is the same on all similar teletype systems.

Because radios are capable of transmitting sound (such as voice or music), eventually people developed ways to send the teletype machine's signals without wires. The audible tones go through the transmitter and receiver, and

Amateurs use other forms of digital radio communications aside from Morse code.

Because radios are capable of transmitting sound (such as voice or music), eventually people developed ways to send the teletype machine's signals without wires.

a terminal at each end decodes the sound into readable text. This is called *radioteletype* or *RTTY* (pronounced RIT-ee). In the early days of RTTY, hams acquired used commercial and military surplus paper-printing terminals and adapted them for Amateur Radio use. Later, video terminals became widely available. Today, RTTY software programs can be built into compact adapters that plug into personal computers.

A related mode is called AMateur Teleprinting Over Radio (AMTOR). This is a more sophisticated form of RTTY that also requires a special terminal device, or a computer connected to an adapter. AMTOR operation involves quickly switching between transmitting and receiving. As the operator types, the AMTOR controller sends characters over the radio in groups of three at a time with an error-checking digit. Then it pauses a moment while the recipient acknowledges the characters as received correctly or asks for a retransmission. The pauses for acknowledgment are only a fraction of a second long, so several groups of characters can be sent each second. AMTOR, then, can be considered a type of error-corrected RTTY.

Packet radio allows reliable computer-to-computer communications. It has become a popular means of relaying routine and emergency messages, and its popularity is still growing.

RTTY and AMTOR are most often operated by a "live" operator, so the person on the other end (and anyone monitoring the frequency) sees your text as you type. RTTY and AMTOR enthusiasts are generally competent typists. If your typing skill is weak, terminal programs for your personal computer let you completely type a line or more of text before it's sent, and you can send whole text files stored on a disk.

Packet: Ham Radio for the Computer Age

Packet radio is a digital mode that lets you use your computer on the air to connect to other hams' computers. It's the same idea as using your computer over the telephone with a modem to connect to other computers, bulletin board systems (BBSs) and information services like CompuServe and Prodigy. Many hams have personal computers in their shacks these days. Having a computer in the shack, however, doesn't mean your station is packet-ready.

In packet radio, a device called a *terminal node controller* (TNC) takes care of two functions. The TNC creates data packets by taking a continuous stream of data coming from your terminal or computer's serial output port and breaking it up into small chunks. Then the TNC's modem converts the packet's digital bits into audible tones. These are sent and received using techniques similar to those used in RTTY and AMTOR.

Packet radio is more sophisticated because each packet of data doesn't have to contain readable text, has much stronger error-correction encoding and tucks more information into each transmission. With RTTY and AMTOR you can only send text messages; in packet, you can send text messages, working program files, graphics and any other computerized information. Because the most common packet operations today use a standardized TNC design, almost any plain data terminal or personal computer with a communications software package will let you operate the TNC.

Amateur packet radio has been on the scene since the 1970s—well before IBM introduced its first personal computer.

Packet communication has been used in wired computer networks in offices, the military and universities for a long time, and amateur packet radio has been on the scene since the 1970s—well before IBM introduced its first personal computer. Amateur Radio operators in the Tucson, Arizona area were instrumental in the development of a TNC kit that thousands of hams used to get on packet radio. All you needed to start communicating via packet was a TNC, radio, a terminal or a personal computer and a communications program.

Thousands of ham packet stations sprang up. Packet was hot stuff. A friend of mine got a new TNC and stayed up all night finding out how it worked and what could be done with it. Even though he was a little sleepy at work

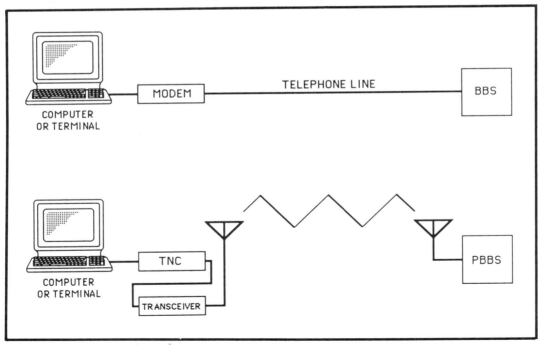

Figure 3—A packet radio bulletin-board system is similar to a telephone-based bulletin board system. One advantage to the packet system is that there are no long-distance phone charges. (On the other hand, you do need an Amateur Radio license to use packet radio.)

the next day, he figured it was worth the effort—he was on the cutting edge.

Packet radio is a useful tool, too. When hurricanes or earthquakes hit, amateur packet networks carry storm and damage updates. Hams use packet radio to exchange more than just messages, however. If you have the right software, you can send vivid computer graphics back and forth. You can send (and receive) complete programs across town or around the world. Hams are experimenting with digitized voice transmission. One day, you may be able to carry on a digitized voice contact with someone around the world via packet radio.

A packet radio message-delivery system is spread across North America. A radio amateur in New York can address a message to a friend in California simply by listing her call sign and that of the packet bulletin board system (PBBS) she uses most often. If he doesn't know the PBBS, he can send it to her call sign and her postal ZIP code, and it will get there faster than it would via the post office!

Hams use packet radio for space communications as well. During a space shuttle mission aboard *Columbia* in December 1990, payload specialist Dr Ron Parise, whose amateur call sign is WA4SIR, left an automatic packet radio "robot" running when he couldn't operate "live." More than 1700 amateurs connected with Parise's spaceborne ham station! Most of the earthbound hams used stations with a 2-meter FM transceiver and a simple vertical whip antenna. Five months later, in April 1991, mission STS-37 made history as an all-ham crew flew shuttle *Atlantis*. They operated on voice and packet radio, and picked up the first amateur television video ever received by humans in space—sent by hams in their homes!

You can monitor packet communications from Amateur Radio satellites without even connecting to them. It's much like "eavesdropping" on teletype feeds from news organizations like the Associated Press, or picking up press news photos or weather-satellite images. Special data controllers provide packet, RTTY, AMTOR, CW, amateur television, two-way fax and weather-fax receiving capabilities.

To get on packet radio, you need a data terminal, teletype machine or a personal computer and communications software program (like the one used with a modem); a TNC; a radio; an antenna and cables to hook them up. A simple TNC costs as little as $50 for a do-it-yourself kit to $150 or more for an assembled unit. Sophisticated multimode communications interfaces can cost several hundred dollars.

Visionaries Wanted

Learning how to connect to the packet network and maximize its usefulness is a challenge. Another area of development is experimenting with higher-speed transmission. Most VHF packet activity is conducted at 1200 bits per second. (1200 bits is about the amount of information in two or three sentences in this book.) Amateurs are experimenting with modems that transfer data at 56,000 bits per second and even 2-4 million bits per second—more than 10,000 pages worth of text!

SPACE COMMUNICATIONS

What's an OSCAR Satellite?

As mentioned in Chapter 2, radio amateurs have had their own satellites in orbit since 1961. Most of them are known as OSCARs, for *O*rbiting *S*atellite *C*arrying *A*mateur *R*adio. You can use the satellites yourself, but communicating through them presents some challenges. For one thing, a regular station isn't moving at speeds as high as 17,000 miles per hour! Tracking a satellite as it speeds overhead takes a particular kind of antenna rotator that can swing the antenna in all directions, like a TV antenna rotator. It must be aimed up and down as well.

If a satellite is moving through the sky at more than

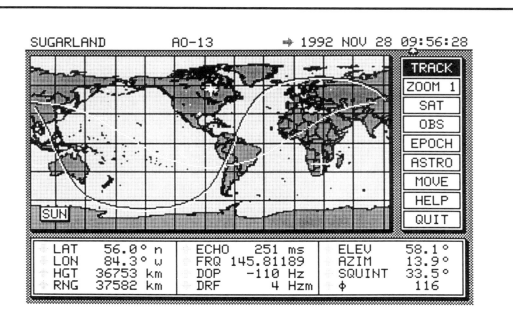

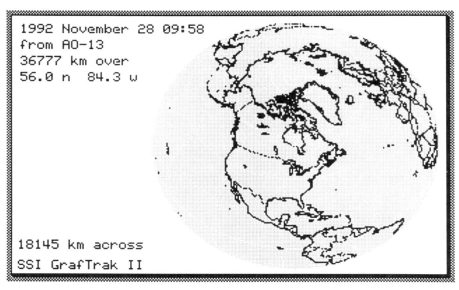

Figure 4—If you want to explore the world of ham radio satellites, you'll need to find out when they're in range of your location. Computer programs make tracking the spacecraft a breeze.

Fuji-OSCAR 20 (the small satellite above the J) was launched by the Japanese space agency in 1990. It's one of a series of amateur satellites that have been providing hams with a direct link to space since 1961.

8J1JBS
on-board Fuji-2/FO-20

Photo by NASDA

20 times the speed of sound, how do you know where to point the antenna and when to move it? Fortunately, you can use satellite locating and tracking programs on a personal computer. AMSAT, the nonprofit Radio Amateur Satellite Corporation, sells excellent tracking programs. Many display the satellites' orbits with outstanding color graphics and include provision for tracking commercial satellites, government weather satellites,

*Get into
satellites and
your shack will
start to
resemble
Mission Control
in Houston.
The antennas
needed for
satellite
operation aren't
exceptionally
large or
expensive.*

manned spacecraft such as the Soviet *Mir* space station, the sun and moon, and even the Hubble Space Telescope.

Get into satellites and your shack will start to resemble Mission Control in Houston. Aside from the colorful computer programs, you can even purchase a device that automatically controls the rotator movement.

The antennas needed for satellite operation aren't exceptionally large or expensive. Depending on which satellites you want to use, you'll need a radio that transmits and receives single-sideband voice, CW or packet on Amateur Radio VHF and UHF frequencies.

Don't get the idea, though, that all satellites require fancy antennas and radios. During space shuttle missions with hams on board, people on the ground worked the shuttle with mobile rigs and hand-held transceivers. You can get started with the satellites using nothing more than the rig you use to work the local repeater. Once you get hooked, though, you'll move up to directional antennas and an SSB/CW rig. To the delight of apartment and condo dwellers faced with rules against large outdoor antennas, there's no need to mount an immense external array of dishes and beams. Many amateurs operate via ham satellites from their basements, using simple antennas.

Moonbounce: 500,000 Miles in Three Seconds

If you were more than a few years old in the late 1960s and early '70s, you probably remember the excitement of listening to the astronauts land on the moon. When TV reporters interviewed them, it took about three seconds for the astronaut to answer. Radio waves are fast—they travel at the speed of light—but it still takes time to go to the moon and back. The round trip is about a half million miles. Think about that.

If you can't go to the moon yourself, why not bounce your voice or Morse code signals off it? *Moonbounce* (abbreviated EME for Earth-Moon-Earth) is a specialized— and fascinating—activity.

As the accompanying story demonstrates, having an extraordinary EME station helps. But some hams with average-size lots and average-size budgets enjoy EME, too. EMEers can contact other moonbounce enthusiasts up to 12,000 miles away—when the moon is right between them.

Meteor Scatter

Hams have found other ways to communicate, as well. One of these is meteor scatter—bouncing signals off the atmosphere as meteors streak by. Each year, the earth's atmosphere is bombarded during 10 or 11 major meteor showers. As a meteor enters the atmosphere, it rapidly heats up to the point of boiling and burns away the minerals it's made of in a process that creates a streak of ions. Radio waves bounce off these ionized areas, and under the proper conditions, hams can make contacts with distant places.

What makes meteor scatter challenging is that the ionization that reflects the signal lasts only a few seconds at a time. The whole contact must be completed in less time than it takes to bite through a candy bar!

BE SEEN ON AMATEUR TV

Some hams think it's better to be seen than heard. To that end, they've developed ways to send television pictures back and forth. Amateur television (ATV) experimenters operate miniature versions of the commercial TV stations we're all familiar with. ATV signals are retransmitted by UHF repeaters (relay stations), usually located on a hill or mountaintop.

For receiving, all you need is a "cable-ready" TV, or a downconverter and a regular TV set. Amateur television is typically transmitted on frequencies your local cable TV company uses for its cable channels 50-55. If you disconnect the coaxial cable from the cable-TV company's wall outlet, hook up a directional antenna for the 430-440 MHz range and point it at a nearby ATV

A PERSONAL EME EXPERIENCE

A Texas ham named Dave Blaschke has one of the most extraordinary EME stations in the world. When we pulled into his driveway, Dave rushed out to the car and said, "Come on in. The moon is about to set." What a strange, but apt, expression.

Dave's station is meticulously neat and orderly. He picked up the microphone, flipped a couple of switches and checked a computer display. Then he keyed the microphone and said, "This is W5UN testing." He let go of the microphone button and we waited. Suddenly a strange and eerie voice came out of the speaker, "This is W5UN testing"—his echo from the moon!

Dave gave each of us turns at the microphone that morning. I don't know when I've had

Some ham radio operators go all out to achieve a goal. Dave Blaschke of Manvel, Texas, uses this impressive antenna array to bounce signals off the moon. (photo by Pete O'Dell)

such a thrill in Amateur Radio as hearing my own voice coming back from the moon. It's an indescribable pleasure. After

repeater, you may well receive clear, colorful amateur television!

To transmit, you'll need a camera (your home camcorder will do fine) and a transmitter that can be assembled or purchased for $300-500. Most UHF ATV stations are limited to a range of 40 miles or so.

If you want to send images around the world, you have to give up the idea of motion. You can send still TV pictures around the world with little more than a each of us had taken a turn, we went outside to look at Dave's antennas. His house is in a rural area and he has several acres of land.

A hundred yards behind the house is his EME antenna. A large "flagpole" stands in the center with all manner of apparatus attached to it. Extending out horizontally from the pole 30 feet in two directions is a boom made of tower sections. Each end of the boom is supported on a stripped-down truck chassis consisting of nothing but the frame and four wheels. More tower sections extend up from each chassis to where they meet and support a boom parallel to the bottom one. That's just the framework to hold the antennas in place. Dave has *48* 2-meter beam antennas mounted to this framework. Two rings of concrete, each about the size of a narrow walkway, encircle the antenna and tower.

To aim the antenna at the moon, Dave activates an electric motor that drives the two chassis around the concrete paths. Another motor tilts part of the framework up and down. You can't buy an installation like this; you need to carefully and meticulously build it yourself.

Because contacts normally cannot be made with stations farther away than just beyond the horizon, you wouldn't expect to contact many overseas ham stations on VHF and UHF frequencies. With his massive EME station, however, Dave is the proud ham who holds the first certificate ever issued by the ARRL for making two-way Amateur Radio contacts on the 2-meter band with stations in 100 countries!

typical HF station. This is a digital mode of operating and you need a converter that turns a still picture into audio tones an SSB transmitter can handle. On the other end, you reverse the process. This process of sending still pictures back and forth is called *slow-scan television (SSTV)*. Some multipurpose digital communications controllers have an SSTV mode built in.

Several space shuttle missions with hams on board have featured SSTV pictures. Tony England, call sign

When shuttle ham/astronaut Tony England flew aboard **Challenger** *in 1985, he relayed images of himself down to hams on earth via slow-scan television.*

WØORE, received an SSTV picture of his wife from a ham ground station in Houston. Tony captured the picture and retransmitted it to the ground station. Even though his wife wasn't able to go into space with him, her picture traveled to the spacecraft and back!

FLYING VIA HAM RADIO

The pilot revs the engine and the plane slowly starts down the runway, building speed. The nose tilts up and the plane is airborne. Shooting skyward, the pilot daringly executes barrel rolls, loops, spins and other aerobatic maneuvers. Then he gingerly brings it home and lands perfectly. Want to do that sort of thing? You can, with ham radio and model airplanes.

There are hams who never operate a radio except for radio-controlled (R/C) scale models. It's something that gets into the blood. Take the plane to its limit. Push it, stretch it, find out just how far it will go. It's beautiful to watch a skilled pilot fly an R/C plane. Groups that fly R/C planes and helicopters are often invited to put on demonstrations at ball games, fairs and other events.

One of the many things you can do with radio outdoors: fly radio-controlled model planes.

If you've always wanted to fly your own model plane, this may be the opportunity you've been waiting for.

DO IT!

We've touched on some of the gadgetry hams use to make their hobby more enjoyable. Reading about it is one thing, but, as hams around the world will tell you, nothing beats actually doing it!

5

Radio Fundamentals

Many hams have had no previous training or experience in electronics. Others are employed as electronics professionals, helping to expand the state of the art. Regardless of their skill level, most hams enjoy knowing how their radios work and how radio waves travel. You'll need to know some radio electronics to qualify for your ham radio license, but a good technical background provides more ham radio enjoyment, as well.

In general, the technical side of ham radio covers:
- what is radio-frequency energy,
- how is it produced,
- how does it get from my station to yours, and
- how does it get turned into a form that can be understood?

These questions span the physics and electronics, with some math thrown in for good measure.

You'll hear hams talking about these subjects fairly often, and for good reason. One of these good reasons is that hams are allowed, and even encouraged, to build their own radio equipment and modify commercial gear.

You might want to replace a component with a new one that will do the job better. Or you may want to build a piece of gear yourself. To do these things well (and safely), you'll need to know something about how radio equipment works.

Knowing how radio waves travel allows you to choose the best frequency band and time of day to contact another ham on the other side of the world, the other side of your state or on top of the next hill. (A band, by the way, is a range of frequencies. CBers have one band to communicate over; hams have many.)

We'll provide a brief look at some radio fundamentals, giving you a taste of what the technical side of ham radio is all about. We'll cover three areas: radio waves and how they travel, the radio-frequency spectrum and antennas.

RADIO WAVES, AND HOW THEY GET FROM MY ANTENNA TO YOURS

If you're going to figure out which band will get you that contact you've been trying to make for weeks now, you'll need to know something about the nature of radio waves and how they travel.

Back around the turn of the century, people thought radio waves were ''vibrations of the aether.'' The aether has fallen into disfavor these days, so we have other explanations. To become a ham you don't need a full understanding of what a radio wave is. (You don't need to be an aeronautical engineer to take a commercial flight to Chicago, either.) But all hams should have some idea of how radio waves get from one place to another.

Picture the ripples that spread out in a pond after a rock is thrown into it. Small waves travel away from where the rock hit the water, in all directions.

Similarly, the *electromagnetic energy* produced in a transmitter and sent to an antenna moves away from the antenna in waves. They travel through the atmosphere at the speed of light, and are picked up by another antenna

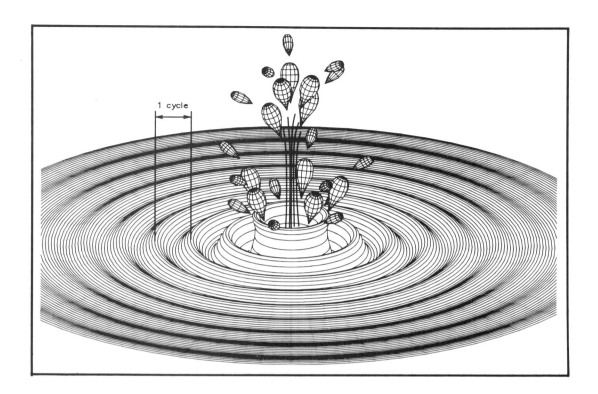

1 cycle

As a rock hits a pond, small waves are created that spread away from the point of impact. Each wave has a peak (high point) and a trough (low point). The complete cycle shown here begins at the peak of one wave and ends at the same point on the next wave.

at some distant point. They are then fed to a receiver, where they are amplified (made stronger) and converted into sound waves.

Frequency

At the pond, you may be able to count how many ripples pass by a certain point in a certain amount of time. Let's say 15 ripples pass by in a span of 15 seconds. This is the same as 1 ripple per second.

As the drawing shows, a ripple can be thought of as a wave *cycle*—a complete wave that first goes up and then down through the starting point, then up again to the starting level. Radio waves consist of cycles, too.

You can't see radio waves, of course, but there are ways of measuring how many cycles can be thought of as passing by a certain point in a certain amount of time.

This is called the *frequency* of the radio wave. Frequency is measured in cycles per second, also known as *hertz.* (The hertz is named for Heinrich Hertz, a 19th century German radio experimenter.)

A radio wave that has 600 thousand cycles in a second has a frequency of 600,000 hertz or 600 kilohertz. (*Kilo* means thousand.) This frequency happens to fall in the range of the AM broadcast band.

The Spectrum

The chart shows where different bands of frequencies lie in the radio-frequency (RF) spectrum. The low-frequency waves at this end of the spectrum were among the first radio waves discovered and used. The Navy still uses this part of the spectrum, because very-low-frequency (VLF) and extremely-low-frequency (ELF) radio signals can pass through water. Navy transmitters that put out millions of watts of power are used to communicate with submerged submarines.

Move up until you reach 540,000 cycles per second (540 kHz). This is the bottom of the AM broadcast band. The other end of the AM broadcast band stretches up to around 1700 kHz (which is the same as 1.7 MHz). MHz

This looks like lopsided piano keys, but it's actually a chart of the electromagnetic spectrum. It shows where the Novice and Technician ham bands (through 1296 MHz) are located in the overall scheme of things. Other ham bands are shown with shorter lines.

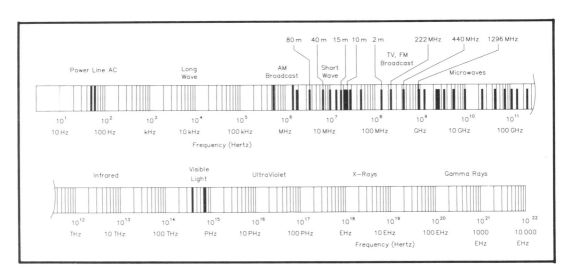

is short for *megahertz*, or millions of hertz. Radio waves at different frequencies travel differently through the air and across the earth.

The lowest Amateur Radio frequencies are just above the AM radio broadcast band. This is the 160-meter ham band. Radio waves at these frequencies *propagate* (travel through the air) about the same way the AM-band signals do. During the daytime, 160-meter signals normally travel less than 100 miles. But at night, they can travel great distances, sometimes halfway around the world.

Name That Band

You may have noticed that each band has two names. It can be called by a *frequency,* such as "3.5 MHz," or it can be called by its *wavelength,* with a name like "80 meters." As the frequency goes up, the wavelength goes down. As the frequency goes down, the wavelength increases. Frequency and wavelength are said to be *inversely related.*

Think of wavelength as the distance the signal travels during one complete cycle. We know that the speed is constant—radio waves travel at the speed of light, 186,000 miles per second or 300,000 kilometers per second.

Either the frequency or the approximate wavelength can be used to identify a band. Hams tend to use *wavelength* for the bands below 222 MHz and *frequency* above this point. For instance, 80 meters (3.5-4.0 MHz), 40 meters (7.00-7.30 MHz), 20 meters (14.000-14.350 MHz), 6 meters (50-54 MHz) and 2 meters (144-148 MHz); but we talk about 222 MHz (1.35 cm), 440 MHz (70 cm) and so on.

Propagation: How Radio Waves Travel

Back to the chart. The 80-meter ham band starts at 3.5 MHz and stretches up to 4.0 MHz. Radio waves in this band travel similarly to those of the 160-meter band. It's a little easier to make long-distance contacts on 80 meters than 160, though.

Radio waves travel around the world by bouncing off a layer of the atmosphere called the ionosphere. *Depending on a number of factors (conditions on the sun, time of day, time of year and frequency of the radio wave, to name a few) radio waves will bounce back to earth, sometimes making several hops. This is how long-distance radio communication works. Radio waves at other frequencies pass through the ionosphere into space.*

The 40-meter band (7-7.3 MHz) is often useful during the day and the night. During the daytime, you may be able to contact stations several hundred miles away. At night, stations roll in from around the world.

The 30-meter band (10 MHz) shares many of the characteristics of the 40-meter band. Because this band is so small, US hams are limited to Morse code (a more efficient use of the spectrum than voice modes).

The major DX bands are 20, 15, 12 and 10 meters. These are most likely to be affected by the *11-year sunspot cycle.* What's the 11-year sunspot cycle? Glad you asked. It's a predictable pattern of the number and extent of dark patches on the surface of the sun. When the sunspot number is high, the sun expels certain types of particles, which cause the upper layer of the earth's atmosphere to become heavily *ionized*—full of charged particles.

Radio waves tend to bounce off these ionized layers, much as light bends as it passes from air into water. A radio wave may make several of these bounces (called *hops*), and thus travel around the earth. This is how you can pick up the BBC on your short-wave radio—and how

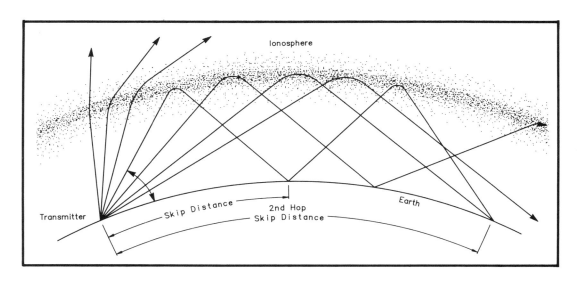

hams in the Northeastern US can communicate with a ham station on a tiny island in the South Pacific.

The 20- and 15-meter bands are more likely to be useful during the daytime hours than at night. Propagation on these bands depends on several factors, however, including sunspot number, the time of year and other atmospheric conditions. During periods of high ionization, you can talk to foreign countries anytime during the day or night on one or both of these bands.

When you go as high in frequency as 50 MHz (the 6-meter amateur band), the ionosphere is rarely energized enough to bounce the signals back to the earth. Before you get a false impression, though, be aware that people have contacted 100 different countries on 6 meters.

Above 50 MHz, RF energy isn't generally reflected by the ionosphere. Other modes of propagation take over, however. You find long-distance "band openings" from 160 meters all the way up through the microwave bands.

We sometimes call the Amateur Radio bands from 80 to 10 meters (3.5 to 29 MHz) the *high-frequency (HF)* bands. The very-high frequency (VHF) ham bands stretch from 6 meters to 1¼ meters (50 MHz to 225 MHz). Next come the ultra-high frequency (UHF) ham bands, from 420 to 1300 MHz. Above 1300 MHz are the microwave ham bands—the largely experimental areas where hams often use exotic modes of communication. Frequencies higher than 300,000 MHz (300 GHz) approach the region of visible light. Above light are cosmic rays, gamma rays, X-rays and other exotic stuff. Hams don't normally venture way up there.

The bands above 144 MHz are primarily local bands; propagation is usually limited to *line-of-sight* (what you see is what you hear). Of course, line-of-sight can cover thousands of miles when you're dealing with orbiting satellites or when you're bouncing radio signals off the moon.

THAT'S IT

This brief glimpse into the technical side of ham radio only scratches the surface. As you gain more experience, you'll want to delve more deeply into these subjects, and others. The ARRL book *Now You're Talking! Discover the World of Ham Radio* covers the electrical theory you'll need to know to earn your first ham license.

6

Your Own
Radio Station

Assembling your first Amateur Radio station is an exciting step toward getting on the air. Once upon a time, the only modes available were Morse code and voice, and commercial equipment wasn't all that common. This served to make life simpler, but it wasn't nearly as much fun as hams have today sending their signals through repeaters, computers and even orbiting satellites.

These days, new hams first have to decide which bands and modes they want to try, and then choose from a wide variety of equipment. This chapter covers the most popular types of transceivers and antennas.

"RIG HERE IS HOME-BREW, OLD MAN"

Hams are fond of using the term "rig" to refer to any *transceiver* (a transmitter and receiver in one unit). This term stems from the early days when most hams built, or "rigged up" their own radio equipment from

Whether it's a wire antenna or a low-power transceiver, nothing beats the satisfaction of putting a station on the air that includes equipment you've put together yourself. In the old days (this photo is from the 1930s), nearly all radio equipment was homemade.

scratch. Today, most hams use store-bought equipment, but if it puts you on the air, it's a rig.

For several years after wireless radio communications became possible, there was no place to buy a factory-made radio set. Besides, it was difficult to find a parts dealer. Amateur Radio began with people who enjoyed putting together a conglomeration of parts and loose ends to make a gadget that would send or receive a signal. Early hams even built their electronic components. They wrapped wire around oatmeal boxes to make coils, and cut up tin cans to make capacitors. Early-day hams were ingenious and resourceful—they had to be.

Today's hams find that it's easy to set up a station and get on the air—if you do your homework and find the equipment that best suits your needs, interests and budget. There are many bands, operating modes and personal preferences to consider. Experienced local amateurs are a good source of guidance. Another is the rest of this chapter.

VHF AND UHF: WAY UP HIGH

Since your operating privileges (the frequencies and modes you're authorized to use) depend on your class of amateur license, your license class also affects which equipment and antennas you'll want to start with.

If you're like many newcomers, you'll start with the codeless Technician license. This class of license allows you to operate on all the VHF, UHF and microwave bands, using all authorized modes. As a code-free Technician, you can use packet radio, satellites, FM repeaters, Morse code and SSB voice. You'll have no trouble finding equipment, new or used, for the popular VHF bands.

Of all the VHF/UHF activities, using local repeaters is the most popular. In fact, the 2-meter FM repeater, which allows you to communicate up to 100 miles or so

This multi-band/multimode VHF transceiver makes a versatile base station. You can use it on three different bands: 2 meters and 430 MHz come standard, and 1240 to 1300 MHz is optional.

using a small, low-powered transceiver, is the most widely used medium in US Amateur Radio. This band is where most Technician and higher-class hams get started. It's a good way to meet other hams on the air. It's also the easiest way to operate.

If you expect to move on to more exotic modes, such as using packet radio over the amateur satellites, trying to contact distant stations using Morse code or bouncing your signals off the moon (hams call this *moonbounce*), consider a multimode transceiver. A typical multimode rig works on FM, CW, single-sideband voice (SSB) and perhaps AM. Multimode, or all-mode, transceivers are more expensive and heavier than FM-only rigs. You won't find multimode hand-held rigs, for example. If your interest lies strictly in earthbound packet operation, pick up an FM-only rig.

NEW OR USED?

Hams are constantly buying and selling radios. There's always a big market for used gear, and it's easy to locate at flea markets or through magazine classified ads, club newsletters, special ham-oriented "shopper" newspapers or via the grapevine.

Although many people start with used gear, it can be risky for a newcomer. If you're good at fixing things, you might not mind a bargain on a radio that needs minor repairs. If you just want to push a button or two and operate, you could be better off buying equipment from hams you know personally or from retail dealers.

If you can afford it, there are advantages to buying your first radio new. You'll be able to pick out the features you want, expect it to work reliably, and be assured of getting it serviced if there's a problem. There's also less chance that your radio will develop problems. Whether you buy new or used, have a knowledgeable ham friend help you pick it out.

FM: THE RELIABLE WORKHORSE

FM rigs are small and fit easily into all types and sizes of vehicles. Modern controls make it simple—and safe—to tune and operate mobile FM transceivers while driving.

A hand-held FM transceiver is small, lightweight and easy to carry, even on a vacation or business trip. The ultimate in "business trip" leisure-time 2-meter FM operation came when astronaut Owen Garriott, W5LFL, carried a 2-meter hand-held transceiver on US space shuttle flight STS-9. For 10 days in late 1983, Garriott electrified the world's Amateur Radio operators with 2-meter FM contacts from space. Hams fortunate enough to listen and call at the right time were able to talk to Garriott from their living room armchairs as the shuttle passed overhead!

FM operation is fun. You can use a small hand-held transceiver to talk directly to amateurs up to several miles away, or extend your range by relaying your signals through special *repeater* stations set up on mountains or

A mobile rig can easily be wired to work in your car, truck, house, boat, office or classroom. With the addition of a power supply, a mobile rig can serve as a base station as well. This compact 2-meter FM transceiver has many features that make operating through a repeater or direct (simplex) more enjoyable.

A good choice for a first rig, a hand-held transceiver can be transported and operated just about anywhere. Many hand-held radios can be interconnected to the phone lines by means of a repeater. This allows local calls to be made from your vehicle.

tall buildings. You can keep in touch with ham friends or make new ones while out for a walk, from poolside or while parked in rush-hour "distressway" traffic. If you need assistance on the road, a call on the local repeater can bring immediate help. Repeaters are the focal point of local clubs and groups that share common interests.

Hand-held radios are also ideal for emergency communications. When there's a hurricane, flood, tornado or blizzard, hams with hand-held radios often help fire, police or Red Cross personnel keep in touch with other emergency workers.

FM transceivers also come in base and mobile versions. Each has some advantages and some drawbacks.

Choosing an FM Radio

Assuming your primary concern is performance on the road, you'll want to look at a mobile transceiver. A new hand-held transceiver will require about the same investment. If you add an optional power amplifier, your hand-held radio will be in the same performance ballpark as a base or mobile rig.

Used radios may cost substantially less, but will have fewer features—and, more than likely, provide fewer years of reliable operation.

The biggest differences between hand-held and mobile transceivers are in size and convenience. The mobile radio takes up more room in the vehicle, but it usually has more standard features than a hand-held, and it's more convenient to use while you're driving.

A mobile rig can be used as a base radio with the addition of a power supply—a device that allows it to plug into a household outlet. The multimode or multiband base transceiver is the best choice for those who want a "serious" radio at their home stations. With a good antenna system (and, perhaps, an amplifier), you can make VHF contacts on SSB and CW and collect "grid squares"—areas of the US and Canada that hams try to contact to earn awards.

It's not an easy decision to choose between purchasing a hand-held, mobile or base transceiver as your first rig. In terms of versatility and portability, the hand-held comes out on top. If you'll do most of your operating while driving, consider a mobile rig. If you want to operate from home using different modes, the base station may best meet your needs.

Think of what you'll do with an FM rig. You'll always have someone to talk to on the way to and from work or school. With a rig in the car, you don't need to be concerned about what you'd do if you broke down on the expressway. Maybe you dream of contacting the next space shuttle mission with hams aboard. The things you can do with an FM rig are limited only by your imagination.

GEARING UP FOR THE HF BANDS

Many beginning hams use the code-free Technician license as a steppingstone to the HF bands. HF is, after all, where the worldwide action is. If your goal is to converse with a farmer in Central America, or a descendant of *HMS Bounty* mutineer Fletcher Christian on a tiny Pacific island, you'll need a license that gives you HF privileges.

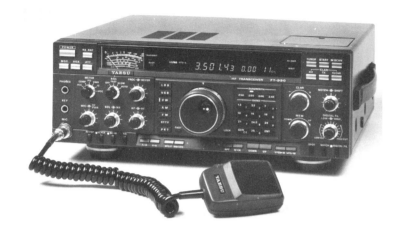

A base-station transceiver offers the most power and features, but can't be carried around and operated as a portable station. These rigs operate on FM, CW, SSB and sometimes AM.

If you choose to earn the Novice or Technician Plus license, you have an even larger range of equipment choices available. Even though you may also want an FM rig to get in on the local action, you'll probably gravitate toward an HF radio.

Unless you come across a creampuff at a flea market or you have solid experience delving into the insides of a transceiver, you'll probably be better off buying a new transceiver. You won't find hand-held transceivers for the bands below 30 MHz, but there are dozens of base station and mobile rigs on the market. Most cover all the HF bands, and many include the 160-meter MF band as well. Some offer optional modules for 6 or 2 meters.

If you're certain you'll be satisfied working only 10 meters, at least for a while, you can pick up a new 10-meter-only rig at a substantial savings over a multiband radio. They're relatively small, and work great as base or mobile rigs.

After you make a handful of CW contacts, you may well discover that you like the code. You'd be surprised how many people have sworn up and down that they were learning Morse code only to pass the exam needed to get on 10 meters. Then they got the hang of it, made a couple of CW contacts and were hooked.

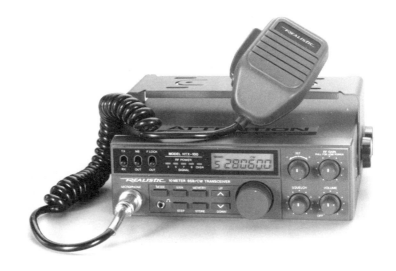

A good way to save money when first starting out is to buy a single-band transceiver, like this one for 10 meters. It operates on two modes, sideband voice and CW.

If you buy an entry-level or midrange HF rig, you'll be able to make worldwide contacts, enter HF contests and compete for awards. Most HF rigs include a general-coverage shortwave receiver. That's an additional bonus when world events are moving faster than the network news teams can keep up with them. Shortwave receivers allow you to tune in to the international broadcast frequencies between the amateur bands and find out what's going on firsthand. Most older rigs on the used market won't receive outside the ham bands.

Even if you have the budget for a top-of-the-line radio, you might be better off starting with one that's less expensive. They're a little "friendlier," because they have fewer knobs, buttons, meters and switches. Most HF transceivers include the basics to make operating pleasurable and easy.

Most hams don't keep gear forever. Every few years, manufacturers trot out more exciting transceivers with fancier features and performance. So if you're looking for a good used radio, you'll have plenty to choose from as hams "trade up." Be sure to take an experienced ham with you when you shop for used gear. If you can, buy

You'll save money by buying a used radio, but you'll need to answer some questions beforehand. Can you get the features you want? Is it in good shape? Is the manual included? Can you get it serviced if necessary? Will you be able to sell it when it's time to move up?

a radio from a local ham. Odds are, he or she will be able to give you some tips on making it perform better and explain any quirks in its operation.

GETTING OUT: A WORD ABOUT ANTENNAS

Now that we've surveyed the types of radios available, we'll take a brief look at antennas.

Hams use several different types of antennas, but the most common are wire, verticals and beams. Whether they're simple or complex, light or heavy, cheap or expensive, antennas are a vital part of all ham radio stations.

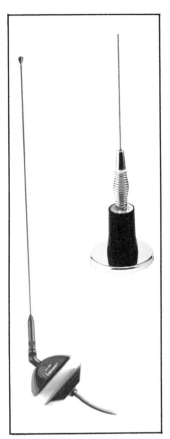

You can drill a hole in your vehicle roof to mount a VHF FM antenna—but most hams mount their antennas temporarily on the roof (mag mount) or on glass. Another choice is a trunk lip mount.

You need an antenna both to send (transmit a signal) and to receive. If you're content to listen to others, you can use an antenna connected to a short-wave radio. What separates hams from short-wave listeners is the ability to enjoy both sending and receiving. To simplify matters, most hams use the same antenna for both.

VHF/UHF Antennas

The type of antenna you connect to your VHF or UHF radio will depend on where it's set up, the bands it covers and your operating preferences. VHF/UHF operators use vertical antennas for FM repeaters, beams (often more than one, for increased performance) for CW and SSB "weak-signal" work, and small "rubber ducks" for hand-held radios. A collapsible 5/8-wavelength vertical antenna improves the performance of hand-held rigs, at the expense of size.

Mobile antennas are convenient to install, and are used on modes of transportation ranging from sailboats and flatbed trucks to mom's station wagon. They can be mounted on top of a vehicle, on a trunk lip or on a windshield. A "mag-mount" antenna is a convenient way to place an antenna on top of a vehicle—some of us don't want to drill a hole in the roof. Some mobile antennas cover two different bands.

Several manufacturers offer good-performing base-station FM antennas. They're often mounted to a mast that attaches to the side of a house.

Beam antennas (the Yagi beam is a common type) are excellent performers and thus are the preferred antennas of those who bounce signals off the moon, and those who compete in contests, where every watt of radiated power counts. The great advantage to beams is that they can be rotated toward a favored direction. Serious VHF enthusiasts use arrays of beams that look impressive and perform the same way.

HF Antennas

Your choice of HF antenna is important to the success of your station. Everyone has a slightly different situation. If you live in a condo or apartment building that prohibits outside antennas, installing a good antenna can be more challenging—but isn't impossible. If you live where antennas are prohibited, contact the American Radio Relay League for advice.

Several types of antennas work well at HF, including the popular and inexpensive wire antenna, the vertical and the beam. Many hams build their own wire and vertical antennas, but many types are sold as well.

Many beginners use a wire antenna on the HF bands. It's simply a length of wire attached to (fed by) the transceiver. It can be attached at one end or in the center. If it's fed in the center, it's a *half-wave dipole*. This type of wire antenna must be a certain length so it will perform well at the particular frequency it's designed for. A single end-fed *random-length wire antenna* can work well on a number of bands, especially when used for receiving.

Many hams use wire antennas for the HF bands. You can get the materials to make a good dipole wire antenna for a few dollars from your local hamfest, radio dealer or Radio Shack outlet. Some dipoles are designed to work on more than one band. The one shown here is perfect for those who live in apartments. The inset shows how a feed line connects to the center of a dipole.

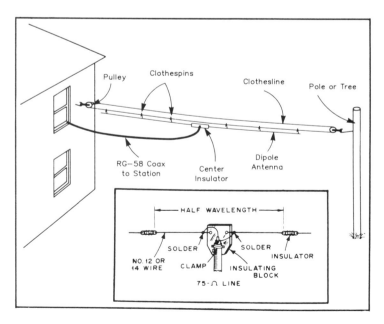

These gents, who live in Iceland, are installing an HF multiband vertical antenna. Verticals can be solid performers, but lack the directionality of beams.

An HF vertical can be designed for a single band or many bands, for mobile use or for a home station. Verticals are *omnidirectional*, which means they transmit in all directions and hear signals coming from all directions. Hams who take their HF radios with them on business trips or vacations often use vertical antennas because they're easy to pack, carry and install.

Beam antennas can be large or small, single-band or multi-band. Experienced hams who have the space and the budget for a tower will want to get the beam antenna up as high as possible.

HF beam antennas have the advantage of being directional—you can point them in the direction you want to reach. In that way, more of your signal is directed where you want it. Likewise, you can hear stations from a particular area better than with an omnidirectional

Let's assume you live in a house on an average-sized lot. Is there a "best" HF antenna for you? Yes and no. You'll do okay with a vertical or a wire, but for the higher HF bands (10-20 meters), you can't beat a steerable beam antenna. Smaller beams can be mounted on a tripod on the roof of your house, while larger ones, like these four, require a tower. Interestingly, the antennas in this photo do not rotate—the tower does!

antenna. If you want to practice your Spanish by talking with some South Americans, for example, you'll want to point your beam antenna (surprise!) to the south.

Beam antennas can be more complicated and expensive than wire and vertical antennas, but many experienced hams will use nothing else because they perform so well.

BUY THE RIGHT GEAR

Be an informed consumer. Spend time with buyer's guides (*CQ* and the ARRL publish such guides—see the "Where to Find It" section at the back of this book). Read product reviews in ham magazines and talk to local amateurs. Figure out which bands, modes and features you want, then buy the rig and antenna that's right for you.

7

Earning Your First Ham License

Before you can transmit on ham radio frequencies, you'll have to pass a test. While that may sound too much like school, there are good reasons for requiring an exam. It's much the same as requiring an exam for a driver's license.

What would the highways be like if anyone who wanted to could drive a car without demonstrating knowledge of the rules of the road? Our highways would be an even bigger mess than they are now!

Why does the FCC require all hams to pass a test? One reason is safety: An understanding of basic electricity and electronics can help prevent serious injury—to you, family members and neighbors as well as your equipment.

In addition to the safety aspect, preparing for your driver's test made you aware of many other useful things all drivers should know—like traffic laws and courtesy. Likewise, preparing for the ham radio exam will make you a better ham radio operator. You'll learn how to maintain your station so it complies with the rules and doesn't interfere with others. You'll learn how to troubleshoot common difficulties—and solve them. In short, you'll learn the "rules of the road"—how to operate your station efficiently and within the letter of the law.

You say you're no electronics whiz? Surprise! Neither are the vast majority of hams. Most aren't electronics professionals; they learned what they needed to know about radio electronics by preparing for the exam—and by practical experience.

WHAT YOU DO (AND DON'T) NEED TO KNOW TO PASS A HAM EXAM

Only part of the amateur exam covers "theory"— what makes a circuit or an antenna work. You'll also need to know FCC rules and on-the-air operating techniques. Many of these are just common sense. If the class of license you're shooting for requires a knowledge of the Morse code, you'll also need to demonstrate that you know the code.

You're *not* going to be tested on whether you can design a station from individual components—you don't need to be an electronics design engineer to be a ham any more than you need to be an automotive engineer to drive a car. You need to know enough radio theory and rules to operate a station without breaking the law. You need to know how to be sure you're transmitting your signals within the ham bands, not where you could interfere with important communications in another radio service. You must operate in a courteous fashion, because you'll share the airwaves with millions of other users.

These concepts are straightforward; most people find they can pass the beginner's exam with little difficulty. Amateur Radio is proud and fortunate to be entrusted

Novice and Technician
——Frequency Allocation Chart——

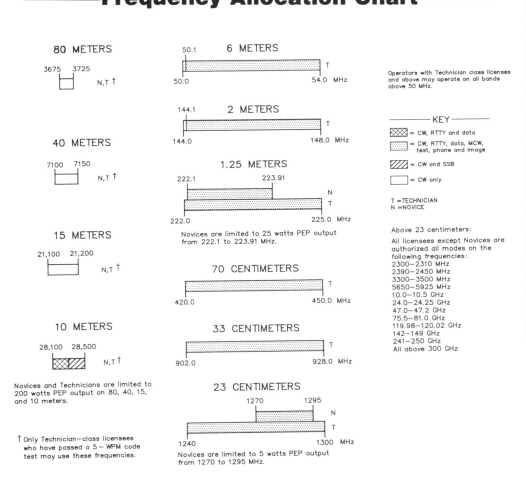

80 METERS

3675 3725

N,T †

6 METERS

50.1

50.0   54.0 MHz

T

Operators with Technician class licenses and above may operate on all bands above 50 MHz.

40 METERS

7100 7150

N,T †

2 METERS

144.1

144.0   148.0 MHz

T

————— KEY —————
⬚ = CW, RTTY and data
▢ = CW, RTTY, data, MCW, test, phone and image
▨ = CW and SSB
▢ = CW only

T = TECHNICIAN
N = NOVICE

15 METERS

21,100 21,200

N,T †

1.25 METERS

222.1   223.91

222.0   225.0 MHz

N
T

Novices are limited to 25 watts PEP output from 222.1 to 223.91 MHz.

Above 23 centimeters:

All licensees except Novices are authorized all modes on the following frequencies:
2300—2310 MHz
2390—2450 MHz
3300—3500 MHz
5650—5925 MHz
10.0—10.5 GHz
24.0—24.25 GHz
47.0—47.2 GHz
75.5—81.0 GHz
119.98—120.02 GHz
142—149 GHz
241—250 GHz
All above 300 GHz

10 METERS

28,100 28,500

N,T †

Novices and Technicians are limited to 200 watts PEP output on 80, 40, 15, and 10 meters.

† Only Technician—class licensees who have passed a 5—WPM code test may use these frequencies.

70 CENTIMETERS

420.0   450.0 MHz

T

33 CENTIMETERS

902.0   928.0 MHz

T

23 CENTIMETERS

1270   1295

1240   1300 MHz

N
T

Novices are limited to 5 watts PEP output from 1270 to 1295 MHz.

with valuable frequency bands for experimentation, public service, spreading goodwill and just having fun. If we abuse these privileges, we could lose them. The airwaves are a limited resource. To paraphrase Will Rogers: "They ain't making no more frequencies these days."

DIFFERENT LICENSE, DIFFERENT PRIVILEGES

The Federal Communications Commission makes the rules for the Amateur Service in the US. Among them is the license structure.

The FCC is no more or less confusing than any other federal agency. For example, the Commission says there are five classes of license, but there are actually six. This came about when the FCC adopted the codeless Technician license in 1991. Instead of creating a completely new class, the FCC set up two kinds of Technician licenses—one with and one without the Morse code endorsement. And so it goes.

So if you pass the Technician written exam, you're a Technician class licensee. If you also pass a Morse code exam, you're a Technician class licensee—but you can also use Morse code on several ham bands. (The "Tech-with-code" license is often called the *Tech Plus* license.)

Ham radio licensing is based on a ladder of license classes and examinations. As you pass each test, you move up the ladder to a higher-class license (often called a ticket*) with more operating privileges.*

Ham radio licensing is based on a ladder of license classes and examinations. As you pass each test, you move up the ladder to a higher-class license (often called a *ticket*) with more operating privileges.

ENTERING THE WONDERFUL WORLD OF HAM RADIO

There are two paths to get into ham radio: The Novice path and the Technician path. (It's possible to start with any class of license, but you'll need to pass all the different exams for each class, individually.)

The *Novice* exam consists of 30 written questions and a 5-word-per-minute (WPM) Morse code test. Novice licensees can use Morse code on parts of the 80, 40 and 15-meter ham bands; voice, code (and other digital modes) on a part of the popular 10-meter band; and all modes on a limited portion of the 222-MHz and 1240-MHz bands. The Novice license is a good, all-around introduction to the joys of ham radio.

The *Technician* exam consists of the same 30-question Novice test and an additional 25-question Technician exam. What do Technicians get out of it? Quite a bit: *all amateur privileges above 30 MHz!*

If a Technician class ham also passes a 5-word-per-minute code test, he or she gains access to parts of the coveted HF bands—the ones hams use to make contacts around the world. This *Technician Plus* license is becoming more and more popular, and for good reason.

AMATEUR LICENSES:
HOW TO EARN THEM AND USE THEM

Class	Code Test	Written Examination	Privileges
Novice	5 WPM	Novice theory and regulations	☐ SSB voice, CW, RTTY and data on 10-meter band ☐ Morse code on three HF bands ☐ All modes on 222-MHz band ☐ All modes on 1270-MHz band
Technician		Novice theory and regulations; Technician-level theory and regulations	☐ All VHF, UHF and microwave frequencies and modes ☐ Technician licensees who have passed a 5-WPM code test also have Novice privileges on 10, 15, 40 and 80 meters
General	13 WPM	Novice theory and regulations; Technician and General theory and regulations	☐ SSB voice on segments of all other HF bands where SSB is authorized ☐ Morse code and most other modes on all amateur bands
Advanced	13 WPM	All lower exam elements, plus Advanced theory	☐ Expanded voice privileges on four HF bands
Extra Class	20 WPM	All lower exam elements plus Extra Class theory	☐ All amateur privileges

To summarize: If you want to use FM repeaters or packet radio, or the many other types of operating available on the VHF and UHF bands, you'll want to earn a codeless Technician license.

If you're also interested in being able to reach other hams across the country and around the world, you'll want the Technician Plus license.

If on the other hand you're most interested in long-distance operating, and aren't that interested in repeaters or packet radio—at least for now—you'll probably want to start with the Novice license.

The table summarizes the advantages of each type of license.

MOVING ON UP

When you're ready to progress from your first license (Novice or Technician), you'll want to earn your General ticket. To move up to General, all you need to do is pass a 13-WPM code exam and an additional 25-question written exam. What's in it for General licensees? Generals can use some part of every amateur band. They can use voice and Morse code to talk to the world on the HF bands (ham radio frequencies below 30 MHz).

The Advanced class license is next. It's based on all the requirements for the General and an additional 50-question written test. With the Advanced ticket comes most of the remaining voice frequencies.

At the top of the ladder is the Extra Class license. This is as far as you can go. It's the ultimate license achievement in Amateur Radio. The Extra Class operator gains access to all of the radio frequencies reserved for hams—after passing an additional 40-question theory-and-rules examination and a 20 WPM code test.

One of the best ways to learn the theory, rules and Morse code (if you're going for a license that requires it) is to take a class.

JOIN A CLUB, TAKE A CLASS

One of the best ways to learn the theory, rules and Morse code (if you're going for a license that requires it) is to take a class. Usually sponsored by a local radio

Many people find a class to be the ideal way to learn about ham radio. You'll have an instructor to answer questions, and you'll get to meet other experienced hams. Your classmates will probably turn out to be ham friends— people you'll stay in touch with on a repeater or over a cup of coffee.

club, classes normally meet one evening a week for 8-12 weeks. Class will probably last 2 or 3 hours. You'll need to buy a book, and code-teaching cassette tapes or a computer program if you're learning Morse code.

That brings up something else you should know: Passing the exams may earn you a license, but ham radio is mostly "on-the-job training." You can cram enough information into your brain to get you through the exams, but your real ham experience comes only after earning your first license and getting some solid on-the-air experience.

HOW TO TAKE AN EXAM

It's easy to take an exam. Novice exams are usually given by two hams involved with your class—or any two hams qualified to give Novice exams. All other exams, including the Technician, fall under the Volunteer Examining (VE) program. Teams of three VEs organize and publicize exam sessions, often in conjunction with hamfests (ham radio flea markets) and ham conventions. As a result, exam opportunities are frequent and widespread.

Taking the exam isn't likely to be the most fun you'll have in ham radio, but it's the only way to earn a license.

VEs are, by definition, *volunteers.* They usually schedule exams for evenings or weekends when most people have time off. Unless you live in a remote rural area, you'll probably have only a short drive to an exam session.

The amateurs who conduct the exams are there because they love ham radio. They want you to do your best on the exams. They're hoping you'll be able to pass. They take their responsibilities seriously, but they do whatever they can to make you at ease. Who knows— you might even *enjoy* it!

There's no "final" exam for an Amateur Radio license. If you don't pass, you can retake it as many times as you need to until you succeed.

There's no "final" exam for an Amateur Radio license. If you don't pass, you can retake it as many times as you need to until you succeed.

Pass or Fail, Keep Trying!

What happens if you fail an exam? Does a ramrod-straight drill instructor come out and say, "Too bad. We thought you had what it takes, but we were wrong. You just aren't good enough to be a ham."

Not a chance! In fact, it's just the opposite. You may even be able to retake the exam (with a different set of questions, of course) the same day! If you elect not to retake the exam that day, and if you pass one part of a

THE ROAD TO YOUR LICENSE GOAL: EXPECT SOME BUMPS AND DETOURS

My wife and I learned some important lessons from our experiences in earning our Extra Class licenses. They're important enough to repeat here for you to learn from.

1) The first and most obvious is that you need to keep taking the exam until you pass. You *will* pass, eventually.

2) Second, some people find one part of the exam difficult, while others don't. Just because your good friend Ralph has trouble with the code doesn't mean you will. There are plenty of people around who breeze through it. Everyone's experience and natural abilities are a little bit different from everyone else's.

3) Tension, whether it's self-induced or brought on from the outside, causes more failures than anything else. If you're tense, learn to control it through relaxation exercises. If you have some knucklebrain in your family who makes you tense, leave him or her home when you go to take the exam.

4) Finally, learning ham radio from your spouse or significant other may not be the easiest way to go about it. Unless you're sure he or she could teach you to drive a car, your best bet is to take a class from a local ham club. If there's no local club offering a class, you can study on your own. It's a good idea, however, to find a ham to call on should you run into a roadblock.

test (the code part of the Novice exam, for example), you're given a Certificate showing the part you passed. It's valid for a year at any other VE session. You're *xpected* to keep taking the exam until you pass!

You might ask, "Wouldn't it be embarrassing if someone found out that I had to take the exam more than once?" Maybe, but I'll let you in on a secret: The author of this book has failed more than a few exams. So has

ELMERS

Spend time around hams and sooner or later you'll hear someone say something like, "Bill was my Elmer." *Elmer?* Who (or what) is an Elmer?

Elmer is a time-honored tradition: an experienced ham who shows a newcomer the ropes. Your Elmer is your teacher, coach, tutor, cheerleader and pal.

Where did the name Elmer come from? Is it some sort of acronym? Actually, Elmer was a real ham. Back in the early days, one fellow who was just starting out wrote a letter to *QST* to publicly thank an established ham who had helped him. Helping new hams along in the hobby is a tradition as old as the hobby itself. The letter-writer's guide's name was Elmer. Through one of those quirks of culture, the name "Elmer" was affectionately adopted to identify any ham who helps another get started.

You can find an Elmer systematically or informally. In some cases, you'll happen to meet an experienced ham at a club meeting or class and he'll decide to show you the ropes. In other cases, clubs set up ways of bringing experienced hams and would-be hams together. This "pairing up" often takes place during a licensing class. It's another good reason to take a class.

An Elmer is there when you don't understand a concept you're learning for the exam. An Elmer is there when you think your code speed isn't coming along fast enough. An Elmer is there when someone offers you a used piece of equipment and you don't know if it's a good deal. An Elmer is there when it's time to put up an antenna. An Elmer is there when it comes time to make the first contact. Your Elmer is there for *you.*

Nothing beats the feeling of accomplishment after you've earned your ham license.

Being a ham isn't about what license class you hold. It's about having fun.

my wife. If you talk to an honest ham (and there are some, just as there are some honest fishermen), you'll find that most have failed at least one exam. (You'll also find a few who have gone from no license to Extra Class in one sitting!)

HAVE FUN WITH YOUR FREQUENCIES

Being a ham isn't about what license class you hold. It's about having fun with the frequencies and modes available to you, whatever they may be.

8

You're on the Air!

Imagine that first minute after you pass your exam. You know that your new, official FCC license will appear in your mailbox in a matter of weeks. What do you do while waiting for your license—and your call sign—to come? Several things.

You'll want to get ready to order some spiffy QSL cards and perhaps call sign license plates for your car. You may want to join a club, and get ready for the next contest or other operating event (if you don't have your license yet, you can watch—or keep the log). These are wonderful things to do, but there's one more little detail—prepare for that day, not too far off now, when you'll be getting on the air!

USING YOUR EARS

Real estate agents say the three most important characteristics of a piece of property are, listed in order of importance: location, location and location. There are three things you should do while waiting for your license to come. They are, listed in order of importance: listen, Listen and LISTEN.

If you have a radio, or can borrow one, spend as much time as you can listening—closely—to the bands and modes you plan to operate on. Listen to what's being said, and how it's being said. It's the best way to learn what to say, and what not to say, once it's your turn to get on the air.

Like a story, most two-way Amateur Radio contacts (called *QSOs*) have an introduction, a body and an ending. The introductory part takes place when two operators establish contact. They begin the body of the conversation when they exchange first names, locations, signal reports and station equipment information. (If the operators already know each other, most of this banter is skipped.) Then one of two things will happen: The conversation will continue, or one of the operators will indicate politely that he wants to move on. Eventually, it's time to end the conversation, and the stations wrap things up—the ending.

SOME DO'S AND DON'TS

If you want to sound good on the air, avoid using phrases used in other radio services, such as police, fire and CB. Hams have a rich store of jargon. You can read about it, but again, the best way to learn it is to listen.

One more thing: Don't take this advice too seriously! All hams foul up once in a while. If you make a blunder, just move on. You'll learn from your mistakes, just as all hams did when they were starting out.

Courtesy

One thing that sets human beings apart from other life forms is self-awareness. We have the ability to place ourselves in the other person's position. We can then ask ourselves if this is how we would like to be treated and then adjust our behavior accordingly.

Suppose you're working in an office and want to call home to see what your spouse would like to have for dinner. Your desk doesn't have a phone, but your neighbor's does. She's talking to someone on her line, but you want the phone. Do you grab the phone out of her hand and make your call? Of course not!

Occasionally, you may hear someone grab a frequency and begin transmitting without listening first. Hams have coined a word for this type of operator: *lid*. It describes a person who engages in any number of undesirable on-the-air operating traits: rudeness, crudeness and thoughtlessness.

The FCC rules state that amateurs must share all frequencies equally and that no one has a "right" or claim to any one of them. No one owns a frequency—although this gets a little more complicated when it comes to repeaters. In that situation, you're using private property (the repeater) and the public airwaves. The point is that hams must share the bands with other hams.

What You Can't (or Shouldn't) Do on the Air

1) *Cause malicious interference*: Deliberately interfering with the communications of other stations is not tolerated by other Amateur Radio operators—or by the FCC. Hams don't put up with such nonsense. They get particularly indignant about unlicensed operators using ham frequencies. Those caught operating illegally will likely be prevented from holding a ham license. In addition, equipment can be confiscated and the person can be fined.

2) *Conduct business of any kind*: As a ham, you can't accept any form of payment for operating an amateur station. In addition, hams aren't allowed to conduct business on the ham airwaves. The FCC defines "business" broadly.

Among the things you can't do over the amateur airwaves are:

• Use a repeater *autopatch* (connection to the telephone system) to dial the local pizza parlor and order a large veggie to go.

• Call someone at your office to tell them you're caught in traffic. (It makes no difference if you work for a not-for-profit organization or government agency; a business is a business.)

It's pretty simple, actually: Using Amateur Radio to conduct any kind of business is illegal over the ham bands.

3) *Play music.* Music of any kind is prohibited on the amateur bands. You can expect some mail from the FCC if you transmit the Theme From Rocky XIV on your local repeater.

4) *Conduct illegal activity.* You can't use the amateur airwaves to assist or promote any activity that's against the law. Nor can you use codes to hide the meaning of your transmissions. (Common amateur abbreviations and Morse code shorthand known as *Q signals* are permitted because most hams understand what they mean.)

5) *Obscenity, indecency and profanity.*

6) *Unidentified transmissions.* If you press the transmit button, you must give your call sign.

READY FOR YOUR FIRST CONTACT?
FIND YOUR ELMER!

One of the most important things an Elmer can do for the newly licensed ham is help him or her get through the first few contacts. Most people rank their leading fears in this order: public speaking, snakes and high places. Although some people never have "mike fright," most of us do at one time or another—especially during those first few contacts. (Okay, if it's a CW contact, it's "key fright"—it's the same feeling.)

Your Elmer can sit beside you and coach you through the contact. He can listen to the other station's transmissions in case you miss them in the heat of the moment. He can tell you how to answer any questions thrown at you. He can even tell you your name if you forget it.

Once you're on the air, you'll find your Elmer to be your ticket to good operating.

(Don't laugh; it happens.) Even if you remember your name, you may forget your call sign or town!

Your Elmer can help you prepare "cheat sheets" to keep nearby with information you're likely to need during a contact. If you're planning to use voice, for instance, it's a good idea to write down your call sign with the proper international phonetics. For instance, WB2D is spoken phonetically as "Whiskey Bravo Two Delta." (These phonetics are available from ARRL Headquarters; see the "Where to Find It" section.) If you plan to operate CW, you might want to have a "typical QSO" written out.

THE VOICE HEARD 'ROUND THE WORLD

Let's look at the way QSOs are conducted on different modes. Assuming you begin with a Novice or Technician Plus license, you'll be able to use voice on the 10-meter band (28.3-28.5 MHz). There can be a lot of activity in that part of the band, so it should be easy to find someone to talk to. Make sure they can hear you and understand you.

If your transmitter is improperly adjusted or tuned, no one is going to hear you. Having an experienced ham around during those first few contacts means you won't have to worry whether your equipment is operating properly.

Here's something else your Elmer can help you with: adjusting your equipment properly. If your transmitter is improperly adjusted or tuned, no one is going to hear you. Having an experienced ham around during those first few contacts means you won't have to worry whether your equipment is operating properly.

Making FM Repeater Contacts

Repeater contacts are different from those you'll make on other modes. Because repeaters are channelized and use FM, with its superior noise rejection and clarity, a repeater contact is almost like talking on the telephone. There's one big difference, however: You must remember to wait until the other station has stopped talking before you start.

A *repeater* is simply a station that automatically retransmits (relays) signals. Unlike a regular station, the repeater uses two frequencies instead of one. The repeater receives on its *input frequency* and transmits on its output frequency.

If you're close to the other station, you won't need to use a repeater. To make a direct contact, you'd use a single frequency—just like SSB contacts on the HF bands. This is called a *simplex* contact.

You'll find that conversations on FM repeaters are more natural and less structured than on other modes, with hams discussing whatever comes up in the conversation. Three or more people often join in a "round-table" conversation.

What will you hear on a repeater? When you want to begin a contact, you'll say:

KB7XYZ listening

If someone wants to talk to you, they'll give you a call:

KB7XYZ from KB7ABC

If no one comes back to you, wait a few minutes and try again.

Repeater users tend to be very mobile, and their operating habits reflect this. They're "in and out" of the

After upgrading his license from Novice to Extra class in less than a year, this ham from North Carolina contacted the Russian **Mir** *space station using a hand-held transceiver and homemade antenna.* **(photo courtesy of Barton Rice)**

system as much as they're in and out of their cars. If you know the call sign of a ham you believe is listening to the frequency, call him. If you're looking for information, just put out a general request. Let common sense be your guide on how to handle a given situation.

Repeaters are costly to own and operate. Transmitters, receivers, antennas, cable, electrical power, telephone service and rent for space on a tower add up to a hefty sum. Because of the expense involved in keeping repeaters on the air, clubs usually own and operate them. If you find that you're using a repeater often, you should join the club that sponsors it.

Making CW Contacts

Most hams are introduced to CW because they want to get a license or upgrade to a higher license class. A lot of hams, however, use only Morse code. They may not have started out that way, but they found that its

A lof of hams use only Morse code.

This young Iowa ham (an Extra Class licensee) can't keep his hands off his key—even while posing for a photo. Newcomer or old-timer, you'll find Morse code superior to voice when conditions are poor. Many hams prefer the rhythms of CW to other modes, regardless of band conditions.

pleasures made it the ideal way to carry on a conversation on the ham bands.

Code is funny that way. When you're new to it, give code a chance. You may find that CW will turn out to be a major source of fun for you. When conditions are bad, or when you're using ham satellites or moonbounce, CW gets through when voice doesn't. It simply gets through noise and interference better than voice.

There's mike fright and there's key fright. Most people are nervous when making their first code contacts. No wonder first CW contacts are so unusual and memorable! Many of the hams you'll find on the CW bands will also be beginners. Experienced code users will slow down to the speed you feel comfortable with. Many amateurs take pride in being a new ham's "first contact." After a while, you'll learn to relax, and your "fist"— the way you send code—will improve miraculously.

A typical CW contact consists of names, call signs, locations, transceiver and antenna, signal report and an ending salutation. If you want to extend the conversation a bit, you can talk about your family, job, school or whatever else interests you.

Morse code is a digital communication language first used in the 19th century.

When you're using Morse code, you're using a digital communication language first used in the 19th century. Those who are fascinated by its rhythmic sounds will no doubt keep it going well into the 21st century.

Connecting to Packet Radio

If you're interested in computers, *packet radio* (described in Chapter 4) may be the mode for you! All it takes is a 2-meter FM transceiver, a computer and a device called a *TNC* (terminal node controller). TNCs are similar to telephone modems—the devices that connect computers to the phone lines. TNCs connect the computer to the transceiver.

What is packet radio? It's computer-to-computer communication over Amateur Radio. You can use packet to enjoy QSOs with other hams in your area. When you establish contact with another station, you *connect* to it. Instead of tapping a key or talking into a microphone, you'll be typing at a keyboard and reading the responses on your monitor. Packet guarantees clear, error-free communications—although it won't correct your typing mistakes!

You can also use packet radio to access your local packet bulletin-board system (or *PBBS*). Packet bulletin boards are warehouses for messages that flow through the worldwide packet network. If you want to know what's happening in Amateur Radio, just check your PBBS. If you'd like to send a message to a ham in another city, state or country, the bulletin board will accept your message and speed it on its way! Do you have a question about your gear—or about anything else related to ham radio? Post a general message on the system and maybe someone will know the answer.

Packet radio is in space, too. Packet satellites make it easy to communicate via orbiting satellite. In addition, the Russian *Mir* space station is packet-active on a regular basis and US space shuttle astronauts frequently operate amateur packet during their missions—in their spare time!

Once you have HF privileges and you're ready to hunt for DX contacts, packet can help! Just look for a *DX PacketCluster* in your area. This is a packet network devoted to DX chasing and contesting. By checking in to a *PacketCluster,* you can find out which DX stations are on the air—and where they are.

Packet has so many features, we could fill the entire book discussing this one subject. It's sufficient to say that packet is a hugely popular mode that's still evolving. If you want to connect your computer to the Amateur Radio world, do it with packet!

THE FUN BEGINS

Before you know it, your FCC amateur license will arrive. Now it's time to order those QSL cards—you'll have plenty of use for them as you make on-the-air friends. You may also want to order a call-sign hat and jacket. Hams are a social bunch, and will step right up and introduce themselves, whether it's at a hamfest or a grocery store!

That's what it's like to be a ham. You can make new friends every time you turn on your radio, and have fun with any number of Amateur Radio activities.

Perhaps you'll be doing it sooner than you think!

This is it—the license that allows you to transmit on the amateur frequencies. Once you have your own license, with its unique call sign, you'll be able to pursue the enjoyable on-the-air activities described in this book.

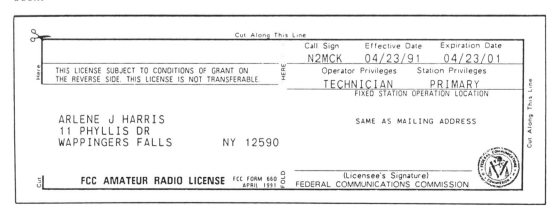

What Next?

Now that you've had an introduction to the wonderful hobby of Amateur Radio, you're probably asking yourself: Okay, what next? Where do I go for more information?

This section is divided into two parts: "Where to Find It" lists companies and organizations that publish Amateur Radio magazines and books, including the material you'll need to earn an Amateur Radio license. The "Dealers" section lists companies that sell Amateur Radio gear.

WHERE TO FIND IT...

Amateur Radio School, KB6MT
2350 Rosalia Dr
Fullerton CA 92635
Phone: 714-990-8442 and 714-990-9622
Amateur Radio audio and video education programs

Ameco Corp.
224 E 2nd St
Mineola NY 11501
Phone: 516-741-5030; *FAX* 516-741-5031
Code-practice oscillators, telegraph keys, amateur accessories, license
   manuals, theory courses, code courses.

AMSAT-NA
PO Box 27
Washington, DC 20044
Phone: 301-589-6062
Membership organization of amateur satellite enthusiasts. Bimonthly
   newsletter, *The AMSAT Journal.* Offers operating aids and volunteer
   assistance. Free information packet.

ARRL, Inc.
225 Main St
Newington CT 06111
Phone: 203-666-1541
Sells direct and through dealers.
Not-for-profit organization of ham radio operators in the US, The American
  Radio Relay League has been the voice of Amateur Radio in the US since
  1914. Publishes *QST* monthly, along with the *National Contest Journal*,
  QEX for experimenters and *The ARRL Letter*, a biweekly newsletter. Also
  publishes licensing books, code-practice tapes and instructor guides. Code
  practice scheduled over Maxim Memorial Station, W1AW. Free beginners
  package. Membership open to everyone with an interest in Amateur Radio.

A.V.C. Innovations
PO Box 20491
Indianapolis IN 46220
Sells direct and through dealers.
Educational products, code courses.

CompuServe
5000 Arlington Center
Box 20212
Columbus OH 43220
Phone: 800-848-8199
HamNet Amateur Radio bulletin board.

CQ Communications, Inc.
76 N Broadway
Hicksville NY 11801
Phone: 516-681-2922; *FAX* 516-681-2926
Publishers of *CQ* magazine, *Popular Communications, ComputerCraft,
  Electronic Servicing & Technology, CQ Amateur Radio Equipment Buyer's
  Guide, CQ Beginner's Guide To Amateur Radio, CQ Videos, CQ Radio
  Amateur* (Spanish CQ) and various Amateur Radio books.

Diamond Systems, Inc.
PO Box 48301
Niles IL 60648
Phone: 312-763-1722
Publisher of Amateur Radio licensing material, computer software, books
  and tapes (Novice through Extra).

GGTE
PO Box 3405
Newport Beach CA 92659
Phone: 714-968-1571
Sells direct and through dealers.
Morse Tutor and Advanced Morse Tutor MS-DOS code software.

HANDI-HAMS
3915 Golden Valley Road
Golden Valley MN 55422
Phone: 612-520-0520

Heath Company
Benton Harbor MI 49022
Phone: 800-253-0570; Technical assistance 616-925-6000
Sells direct.
Amateur Radio courses and home-study electronics courses

Master Publishing, Inc.
14 Canyon Creek Village
Mail Stop 31
Richardson TX 75080
Phone: 214-907-8938
Gordon West study books through dealers and Radio Shack stores.

Media Mentors
PO Box 131646
Staten Island NY 10313-0006
Phone: 718-983-1416; *FAX* 718-983-1416
Educational curriculum, video, computer disks, code keys, teacher's manual.

MFJ Enterprises, Inc
PO Box 494
Mississippi State MS 39762
Phone: 800-647-1800, 601-323-5869
Code oscillators, books, licensing, code and theory programs.

Milestone Technologies
3551 S Monaco Pkwy, Suite 223
Denver Co 80237
Phone: 303-752-3382; Technical info: 303-752-3382
Sells direct and through dealers.
Morse code training and log-keeping software for MS DOS personal
    computers.

PASS Publishing
Box 570
Stony Brook NY 11790
Phone: 516-584-8164; *FAX:* 516-584-9409
Sells direct.
Hypnosis (visualization and autosuggestion) tapes to improve Morse code
    skills and performance on ham code and theory exams

QSO Software
208 Partridge Way
Kennett Square PA 19348
Phone: 215-347-2109
IBM and Macintosh tutorial software for Amateur Radio licenses, control
    software for Kenwood rigs

Radio Amateur Callbook, Inc.
1695 Oak St
PO Box 2013
Lakewood NJ 08701
Phone: 908-905-2961; BBS: 908-905-3074
Sells direct and through dealers.
North American and International Callbooks, Gordon West Radio School,
  maps and atlas.

RadioScan Magazine
8250 NW 27 St, Suite 301
Miami FL 33122
Phone: 305-594-7734; *FAX:* 305-594-7677
Sells direct and through dealers.
Monthly magazine for ham radio and computers. Spanish version available.

Radio Shack
1500 One Tandy Center
Fort Worth TX 76102
Phone: 817-390-3011
Over 7000 stores in the US. Computers, scanners, antennas, transceivers,
  coax, plugs, jacks, parts and supplies.

StarGate Systems
1215 Beechwood Dr
Munford TN 38058
Phone: 901-837-3518; Tech: 901-837-3518
Sells direct.
IBM (MS-DOS) compatible software, mostly public domain and shareware,
  also feature "Build a Disk."

Tucson Amateur Packet Radio (TAPR)
Box 12925
Tucson AZ 85732
Phone: 602-749-9479; *FAX*: 602-749-5636
Nonprofit scientific and educational organization committed to research and
  development in the field of digital communications for hams. Provides
  kits for upgrading packet radio equipment and distributes computer share-
  ware programs related to packet radio.

VIS Study Cards
PO Box 16646
Hattiesburg MS 39404-6646
Phone: 601-261-2601
Sells direct and through dealers.
Publishes complete line of compact flash cards for Novice through Extra
  theory.

W5YI Group
PO Box 565101
Dallas TX 75356
Phone: 817-461-6443
Sells direct and through dealers.
License study guides and practice tapes, *The W5YI Report* newsletter.

WGE Publishing Inc.
WGE Center
Peterborough NH 03458-1194
Phone: 800-225-5083, 603-525-4201; *FAX:* 603-525-4423
Sells direct and through dealers.
Publishes monthly *73* magazine.

Gordon West Radio School
2414 College Dr
Costa Mesa, CA 92626
Phone: 714-549-5000, *FAX:* 714-434-0666
Licensing seminars, local and national classes
Amateur Radio code and theory cassette tapes, radio theory books, and video
  cassette instructor training materials.

Westlink Report
28221 Stanley Ct
Canyon Country CA 91351
Phone: 800-HAM-7303, 805-251-5558
Sells direct and through dealers.
Bi-weekly newsletter.

Worldradio, Inc.
2120 28th St
Sacramento CA 95818
Phone: 916-457-3655
Sells direct and through dealers.
Publishes *Worldradio,* a monthly Amateur Radio magazine.

Zihua Software
PO Box 51601
Pacific Grove CA 93950
Phone: 408-372-0155
Sells direct and through dealers.
Morse code educational software for the Macintosh computer.

————————————————————————————

This list of dealers is taken from the *CQ Buyers'
Guide*, and was up-to-date when this book was published.
Note that toll-free "800" telephone numbers should be
used only to place an order for a purchase, not to gather
general information, for repairs, etc. This reduces the time
that other callers have to wait (it might be you!) for someone to take their order.

A

A-B-C Communications
17550 15th Avenue NE, Seattle, WA 98108.
Phone: 206 364-8300. Sells via showroom.

A.S.A.
PO Box 30162, Charlotte, NC 28230. Phone: 704-358-4815,
   1-800-722-2681. Sells via mail order.

A-TECH Electronics
2210 W. Magnolia Blvd., Burbank, CA 91506.
Phone: 818 845-9203, FAX: 818 846-2298. Sells via showroom
   and mail order.

AXM Incorporated
11791 Loara St., Suite B, Garden Grove, CA 92640-2321. Phone:
   714 638-8807, FAX: 714 638-9556. Sells via mail order and
   direct.

Ack Radio Supply Company
3101 4th Avenue, South Birmingham, AL 35233. Phone: 205
   322-0588, 800-338-4218, FAX: 205-322-0580. Sells via
   showroom and mail order. Branch: 554 Deering Road, Atlanta,
   GA 30367. Phone: 404 351-6340, FAX: 404 351-1879.

ALL Band Radio Products (See Canada Listings).

All Electronics
PO Box 567, Van Nuys, CA 91408. Phone: 800 826-5432,
Tech: 818 904-0524, FAX: 818 781-2653. Sells via showroom and
   mail order. Full line of parts, rechargeable batteries,
   capacitors, semi-conductors, fuses, fasteners, lamps, opto-
   electronics, switches, etc. Branches: 905 S. Vermont Avenue,
   Los Angeles, CA 90006. Phone: 213 380-8000, 14928 Oxnard
   Street, Van Nuys, CA 91411. Phone: 818 997-1806.

Amateur & Advanced Communications
3208 Concord Pike, Rt 202, Wilmington, DE 19803.
   Phone: 302 478-2757. Sells via showroom and mail order.

Amateur Communications, Etc.
263 Mink, San Antonio, TX 78213-3949. Phone: 512 733-0334,
   512 734-793-7794. Sells via showroom and mail order.

Amateur Electronic Supply, Inc.
5710 W. Good Hope Road, Milwaukee, WI 53223.
Phone: 800 558-0411, 414 358-0333. Sells via showroom and
mail order. Used gear.

Amateur Electronic Supply, Inc.
28940 Euclid Avenue, Wickliffe, OH 44092.
Phone: 800 321-3594, 216 585-7388. Sells via showroom and
mail order. Used gear.

Amateur Electronic Supply, Inc.
621 Commonwealth Avenue, Orlando, FL 32803.
Phone: 800 327-1917, 407 894-3238. Sells via showroom and
mail order. Used gear.

Amateur Electronic Supply, Inc.
1898 Drew Street, Clearwater, FL 34625. Phone: 813 461-4267.
Sells via showroom and mail order. Used gear.

Amateur Electronic Supply, Inc.
1072 N. Rancho Drive, Las Vegas, NV 89106. Phone: 800
634-6227, 702 647-3114, FAX: 702 647-3412. Sells via
showroom and mail order.

Amateur Radio Supply Co.
5963 Corson Avenue S., #140, Seattle, WA 98108-2707. Phone:
206 767-3222, 800 457-2277. Sells via showroom and mail
order.

Amateur Radio Team Of Spokane S.
25 Girard, Spokane, WA 99212. Phone: 509 928-3073.

American Electronics
Box 301, 173 East Broadway, Greenwood, IN 46142.
Phone: 317 888-7265, 800 872-1373, FAX: 317 888-7368. Sells
via showroom and phone/mail order to wholesale dealers.

Antenna Service
165 Olympia Street, Pittsburgh, PA 15211. Phone: 412 431-5171.

Antennas Etc.
PO Box 4215, Andover, MA 01810-4215. Phone: 508 475-7831,
508 975-2711, FAX: 508 474-8949. Sells direct and through
dealers.

Antique Electronic Supply
1221 South Maple, Tempe, AZ 85283. Phone: 602 820-5411,
FAX: 602 820-4643. Sells via mail order. Tubes and parts for
tube equipment.

Arnold Company
PO Box 512, Commerce, TX 75428. Phone: 214 395-2922, FAX:
214 395-2340. Sells via showroom and mail order.

Associated Radio Comm.
PO Box 4327, 8012 Conser, Overland Park, KS 66204. Phone: 913 381-5900, FAX: 913 648-3020. Sells via showroom and mail order.

Atlantic Ham Radio, Ltd. (See Canada Listings).

Austin Amateur Radio Supply
5325 North IH-35, Austin, TX 78723. Phone: 512 454-2994, 800 423-2604. Sells via showroom and mail order.

B

B.C. Communications, Inc., The
211 Bldg-Depot Road, Huntington Station, NY 11746. Phone: 516 549-8833, 516 549-1277, 800 924-9884, FAX: 516 549-8820. Sells via showroom and mail order.

BCD Electro
PO Box 450207, Garland, TX 75045-0207. Phone: 214 343-1770, FAX: 214 343-1854. Sells via mail order.

Barry Electronics, Corp.
512 Broadway, New York, NY 10012. Phone: 212 925-7000, FAX: 212 925-7001. Sells via showroom and mail order.

Base Station, Inc.
1839 East Street, Concord, CA 94520. Phone: 415 685-7388. Sells via showroom and mail order.

Boucher Electronics
4813 Lexington Avenue, Erie, PA 26509.

Broadcast Systems, Co.
PO Box 3536, Albuquerque, NM 87190-3536. Phone: 800 777-1412, 505 884-8235. Sells via showroom and mail order.

Burghardt Amateur Center, Inc.
PO Box 73, 182 N. Maple St., Watertown, SD 57201-0073. Phone: 605 886-7314, 605 886-7382, FAX: 605 886-3444. Sells via showroom and mail order.

Burk Electronics
35 N. Kensington, LaGrange, IL 60525. Phone: 708 482-9310. Sells via mail order.

C

C.A.T.S.
7368 S.R. 105, Pemberville, OH 43450. Phone: 419 352-4465. Sells via mail order. Parts and service on all American rotators, cable, wire and rotator accessories.

CBC International, Inc.
PO Box 31500, Phoenix, AZ 85046. FAX: 602 996-8700. Sells via mail order. CB-to-ham radio modification, plans and hardware; FM conversion kits, books, plans, accessories.

C-Comm
6115 15th Avenue NW, Seattle, WA 98107. Phone: 800 426-6528,
206 784-7337, FAX: 206 784-0541. Sells via showroom and
mail order.

COMB
1405 N. Xenium Lane, Minneapolis, MN 55441.
Phone: 612 557-8000, 800 328-0609.

C & S Sales
1245 Rosewood Avenue, Deerfield, IL 60015.
Phone: 800 292-7711, 708 541-0710, FAX: 708 520-0085. Sells
via mail order.

Cable X-Perts, Inc.
113 McHenry Road, Suite 240, Buffalo Grove, IL 60089-1797.
Phone: 708 506-1811, 800 828-3340, FAX: 708 506-1970. Sells
via mail order and hamfests.

Century 21 Communication (See Canada Listings.)

Colorado Comm Center
525 E. 70th Ave, Suite 1W, Denver, CO 80229. Phone: 800
227-7373, 303 288-7373. Sells via showroom and mail order.

Communications City
175 SE 3rd Avenue, Miami, FL 33131. Phone: 305 579-9709,
FAX: 305 570-9808.

Communications Data Corporation
1051 Main Street, St. Joseph, MI 49085. Phone: 616 982-0404,
FAX: 616 982-0433. Sells via showroom and mail order.

Communications Electronics Inc.
PO Box 1045, Ann Arbor, MI 48106-1045.
Phone: 800 USA-SCAN, 313 973-8888. Scanners, Radar
detectors, amateur radio telephones, shortwave, satellite two-
way radio.

Comm-Pute, Inc.
1057 East 2100 South, Salt Lake City, UT 84106.
Phone: 801 467-8873, 800 942-8873. Sells via showroom and
mail-order.

Computeradio
Box 282, Pine Brook, NJ 07058. Phone: 201 227-0712,
FAX: 201 808 1970. Sells via showroom and mail order.
Computers, desktop publishing systems and software, foreign
language systems, laser master typesetters, consulting design,
engineering.

Consolidated Electronics
705 Watervliet Avenue, Dayton, OH 45420.
Phone: 800 543-3568, 513 252-5662, FAX: 513 252-4066;
Telex: 288-299.

Contact East, Inc.
335 Willow St. So., North Andover, MA 01845.
   Phone: 508 682-2000, 800 225-5370, FAX: 508 688-7829. Sells direct via phone and mail order. Major lines: Precision hand tools and tool kits.

D

DANDYS
120 North Washington, Wellington, KS 67152.
   Phone: 316 326-6314. Sells via showroom, hamfests, and mail order.

De La Hunt Electronics
Highway 34E, Park Rapids, MN 56470. Phone: 218 732-3306.

Delaware Amateur Supply
71 Meadow Road, New Castle, DE 19720. Phone: 302 328-7728, 800 441-7008. Sells via showroom and mail order.

Dentronics
6102 Deland Road, Flushing, MI 48433. Phone: 313 659-1776, 800 722-5488, FAX: 313 659-1280. Sells via showroom and mail order.

Doc's Communications
702 Chickamauga Avenue, Rossville, GA 30741.
   Phone: 404 866-2302, 404 861-5610, FAX: 404 866-6113. Sells via showroom and mail order.

E

Easytech Inc.
2917 Bayview Drive, Fremont, CA 94538. Phone: 510 770-2300, 800 528-4044, FAX: 510 770-2346, 800 582-1255. Sells via mail order. ICs and components, cabinets, tools, test equipment, specialty and reference books, Amateur Radio accessories, connectors, adapters, RF power modules, power supplies, export sales.

El Original Electronics
1257 East Levee, Brownsville, TX 78520. Phone: 512 546-9846, 512 542-8507. Sells via showroom.

Electro-Comm
961 E. 65th Street, Tacoma, WA 98404-2237.
   Phone: 206 473-9225, 800 821-9150, FAX: 206 473-9225. Sells via showroom and mail order.

Electronic Center, Inc.
2809 Ross Avenue, Dallas, TX 75201. Phone: 214 969-1936. Sells via showroom, hamfest, and mail order.

Electronic Equipment Bank (EEB)
323 Mill Street, NW Vienna, VA 22180. Phone: 800 368-3270, 703 938-3350, FAX: 703 938-6911. Sells via showroom and mail order.

Electronic Engineering
PO Box 337, Barrington, IL 60011. Phone: 708 540-1106.

Electronic International Services
11204 Spur Wheel Lane, Rockville, MD 20854. Phone: 301
983-3033.

Electronic Module
601 North, Turner Hobbs, MN 88240. Phone: 800 688-0073,
919 791-8885.

Electronic Specialists
Cinema Square Shopping Center, 3830 Oleander Drive,
Wilmington, NC 28403. Phone: 800 688-0073, 919 791-8885.
Sells via showroom and mail order.

Eli's Amateur Radio, Inc.
2513 SW Ninth Avenue, Ft. Lauderdale, FL 33315.
Phone: 305 525-0103, 305 944-3383, 800 780-0103,
FAX: 305 944-3383. Sells via showroom and mail order.

Erickson Communications
5456 N. Milwaukee Avenue, Chicago, IL 60630.
Phone: 800 621-5802, 312 631-5181. Sells via showroom and
mail order.

F & M Electronics
153 W. Railroad Street, N. Pelham, GA 31779-1201. Sells via
showroom and mail order.

Fair Radio Sales
1016 E. Eureka Street, Lima, OH 45802. Phone: 419 227-6573,
419 223-2196, FAX: 419 227-1313. Sells via showroom and
mail order. Military surplus, receivers, test equipment, vacuum
tubes, electronic parts.

First Call Communications (FCC)
3 Chestnut Street, Suffern, NY 10901. Phone: 800 426-8693,
914 357-7339. Towers.

G

Gilfer Shortwave
52 Park Avenue, Park Ridge, NJ 07656. Phone: 201 391-7887,
800 GILFER 1. Sells via showroom and mail order.

H

H.R. Electronics
722-24 Evanston Avenue, Muskegon, MI 49442.
Phone: 616 722-2246. Sells via showroom and mail order.

H.S.C. Electronic Supply
6819 Redwood Drive, Cotati, CA 94931. Phone: 707 792-2277,
FAX: 707 792-0146 BBS: 707 527-7734 8N1 (300-2400).

Hal-tronix, Inc.
12671 Dix Toledo, Hwy Southgate, MI 48195.
  Phone: 313 281-7773. Sells via showroom and mail order.

Ham Buerger, Inc.
417 Davisville Road, Willow Grove, PA 19090.
  Phone: 215 659-5900, FAX: 215 659-5902. Sells via showroom
  and mail order.

Ham Radio Outlet
933 N. Euclid Street, Anaheim, CA 92801. Phone: 800 854-6046,
  Local: 714 533-7373, FAX: 714 533-9485. Sells via showroom
  and mail order.

Ham Radio Outlet
6071 Buford Highway, Atlanta, GA 30340. Phone: 404 263-0700,
  800 444-7927, FAX: 404 263-9548. Sells via showroom and
  mail order.

Ham Radio Outlet
8400 E. Iliff Avenue #9, Denver, CO 80231.
  Phone: 303 745-7373, 800 444-9476, FAX: 303 745-7394. Sells
  via showroom and mail order.

Ham Radio Outlet
2210 Livingston Street, Oakland, CA 94606.
  Phone: 510 534-5757, 800 854-6046, FAX: 510 534-0729. Sells
  via showroom and mail order.

Ham Radio Outlet
11705 SW Pacific Highway Suite Z, Portland, OR 97223.
  Phone: 503 598-0555, 800 854-6046, FAX: 503 684-0469. Sells
  via show-room and mail order.

Ham Radio Outlet
1702 W. Camelback Road, Suite 4, Phoenix, AZ 85015.
  Phone: 602 242-3515, 800 444-9476, FAX: 602 242-3481. Sells
  via showroom and mail order.

Ham Radio Outlet
224 N. Broadway, Salem, NH 03079. Phone: 603 898-3750,
  800 444-0047, FAX: 603 898-1041. Sells via showroom and
  mail order.

Ham Radio Outlet
5375 Kearny Villa Road, San Diego, CA 92123.
  Phone: 619 560-4900, 800 854-6046, FAX: 619 560-1705. Sells
  via showroom and mail order.

Ham Radio Outlet
510 Lawrence Expwy #102, Sunnyvale, CA 94086.
  Phone: 408 736-9496, 800 854-6046, FAX: 408 736-9499. Sells
  via showroom and mail order.

Ham Radio Outlet
6265 Sepulveda Blvd., Van Nuys, CA 91411.
   Phone: 818 988-2212, 800 854-6046, FAX: 818 988-4326. Sells
   via showroom and mail order.

Ham Radio Outlet
14803 Build America Drive, Bldg B, Woodbridge, VA 22191.
   Phone: 800 444-4799, 703 643-1063, FAX: 703 494-3679. Sells
   via showroom and mail order.

Ham Radio Toy Store, Inc.
117 West Wesley Street, Wheaton, IL 60187.
   Phone: 708 668-9577. Sells via showroom and mail order.

Ham Station, Inc.
220 N. Fulton Avenue, Evansville, IN 47710.
   Phone: 800 729-4373, 812 422-0231, FAX: 812 422-4253. Sells
   via showroom and mail order.

Ham Store, The
5707A Mobud, San Antonio, TX 78238. Phone: 800 344-3144,
   FAX: 512 647-8007. Sells via showroom and mail order.

Hamtronics/Trevose
4033 Brownsville Road, Trevose, PA 19047.
   Phone: 215 357-1400, 800 426-2820, FAX: 215 355-8958. Sells
   via showroom and mail order.

Hardin Electronics
5635 E. Rosedale Street, Ft. Worth, TX 76112.
   Phone: 817 429-9761, 800 433-3203, FAX: 817 457-2429. Sells
   via showroom and mail order.

Hatry Electronics
500 Ledyard Street, Hartford, CT 06114. Phone: 203 296-1881,
   FAX: 203 296-7110. Sells via showroom and mail order.

Heaster, Inc., Harold
84 North Timber Creek Road, Ormond Beach, FL 32174.
   Phone: 904 672-2878. Sells via showroom and mail order.

Henry Radio Inc.
2050 Bundy Drive, Los Angeles, CA 90025. Phone: 800 877-7979,
   213 820-1234, FAX: 213 826-7790. Sells via showroom.

Hialeah Communications
630 E. 9th Street, Hialeah, FL 33010. Phone: 305 885-9929,
   FAX: 305 888-8768. Sells via showroom and mail order.

Hirsch Sales Corporation
219 California Drive, Williamsville, NY 14221.
   Phone: 716 632-1189, FAX: 716 632-6304. Sells via showroom
   and mail order.

Hobbytronique Inc. (See Canada Listings).

Honolulu Electronics
819 Keeaumoku Street, Honolulu, HI 96814.
   Phone: 808 949-5564, 808 949-5565, FAX: 808 949-1209. Sells
   via showroom and mail order.

Hooper Electronics
1702 Pass Road, Biloxi, MS 39531. Phone: 601 432-1100, 601
   432-0584.

I

International Radio & Computer
3804 South US #1, Fort Pierce, FL 34982-6620.
   Phone: 407 489-5609, FAX 407 464-6386. Sells via showroom
   and mail order. High-performance crystal filters for popular
   transceivers, newsletters, enhancement kits for popular
   transceivers.

International Radio Exchange
19 Ann Boulevard, Spring Valley, NY 10977. Phone: 800
   321-1069, 914 356-4054.

International Radio Systems
5001 NW 72nd Avenue, Miami, FL 33166-5622. Phone: 305
   594-4313, FAX: 305 477-4449.

J

J.R.S. Distributors, Inc.
646 W. Market Street, York, PA 17404. Phone: 717 854-8624,
   FAX: 717 854-8624. Sells via showroom and mail order.

Jones & Associates, Marlin P.
PO Box 12685, Lake Park, FL 33403-0685. Phone: 407 848-8236,
   FAX: 407-844-8764. Sells via mail order catalogs. Connectors,
   fans, motors, power supplies, meters, switches, knobs, LED's,
   semiconductors, tools, relays, lens, lasers, valves, NiCds.

Juneau Electronics
8111 Glacier Hwy, Juneau, AK 99801-8035.
   Phone: 907 586-2260.

Jun's Electronics
5563 Sepulveda Blvd., Culver City, CA 90230.
   Phone: 213 390-8003, 800 882-1343. Sells via showroom and
   mail order.

K

K-Com
Box 82, Randolph, OH 44265. Phone: 216 325-2110. Telephone
   Interface Filters.

KC2NB Tower Service
RD#1 Box 54A, Elizabeth Ave., Somerset, NJ 08873-9764.
   Phone: 908 873-2198, FAX: 908 873-3441. Sells via mail order.

KJI Electronics
66 Skytop Road, Cedar Grove, NJ 07009. Phone: 201 239-4389.
   Sells via mailorder and hamfests.

L

LaCue Communications, Inc.
132 Village Street, Johnstown, PA 15902. Phone: 814 536-5500.

LaRue Electronics
1112 Grandview Street, Scranton, PA 18509. Phone: 717
   343-2124. Sells via showroom and mail order.

Lentini Communications
21 Garfield Street, Newington, CT 06111. Phone: 203 666-6227,
   800 666-0908. Sells via showroom and mail order.

Lindsay Publications, Inc.
PO Box 12, Bradley, IL 60915. Phone: 815 468-3668.

Litsche, N.E.
PO Box 191, Canandaigua, NY 14424-0191.
   Phone: 716 394-9099, 716 394-0148. Sells via mail order.
   Military surplus test equipment and radios.

Longs Electronics
2630 South Fifth Avenue, Irondale, AL 35210.
   Phone: 800 633-3410, 800 292-8668. Sells via showroom and
   direct mail.

M

MacFarlane Electronic, Ltd., H.C.
(See Canada Listings).

Madison Electronics
12310 Zavalla Street, Houston, TX 77085. Phone: 713 729-7300,
   800 231-3057, FAX: 713 358-0051. Sells via mail order.

Maryland Radio Center
8576 Laureldale Drive, Laurel, MD 20707. Phone: 301 725-1212,
   800 447-7489, FAX: 301 725-1198. Sells via showroom.

McCarthy, N6CIO, Loraine
2775 Mesa Verde Dr. E., Ste E101, Costa Mesa, CA 92626.
   Phone: 714 979-CODE. Sells via showroom and mail order.
   Training materials.

McClaran Sales, Inc.
PO Box 2513, Vero Beach, FL 32961. Phone: 407 567-8224,
   800 331-6186. Sells via mail order.

Memphis Amateur Electronics, Inc.
1465 Wells Station Road, Memphis, TN 38108.
   Phone: 800 238-6168, 901 683-9125. Sells via showroom and
   mail order.

Miami Radio Center Corp.
5590 W. Flagler Street, Miami, FL 33134. Phone: 305 264-8406.
Sells via showroom.

Michigan Radio
23040 Schoenherr, Warren, MI 48089. Phone: 313 771-4711,
800 878-4266 Service: 313 771-4712, FAX 313 771-6546. Sells
via showroom and mail order.

Mike's Electronics
1001 NW 52nd Street, Ft. Lauderdale, FL 33309.
Phone: 305 491-7110, 800 427-3066. Sells via showroom and
mail order.

Mouser Electronics
2401 Hwy. 287, North Mansfield, TX 76063.
Phone: 800 346-6873, FAX: 817 483-0931.

N4EDQ Amateur Radio Sales & Service
4400 Hwy. 19A, Suite 10, Mt. Dora, FL 32757.
Phone: 904 589-0222. Sells via showroom and mail order.

N & G Distributing
1950 NW 94th Avenue, Miami, FL 33126. Phone: 303 592-9685.

National Tower Company
PO Box 15417, Shawnee Mission, KS 66215.
Phone: 913 888-8864.

North Olmsted Amateur Radio Depot
29462 Lorain Road, N. Olmsted, OH 44070.
Phone: 216 777-9460. Sells through showroom and mail order.
Accepts trade-ins; does repairs.

Ocean State Electronics
279 High Street, Westerly, RI 02891. Phone: 401 596-3080.

Oklahoma Comm Center
9500 Cedar Lake Avenue, Suite 100, Oklahoma City, OK 73114.
Phone: 405 478-2866, 800 765-4267, FAX: 405 478-4202 Sells
via showroom and mail order.

Omar Electronics
2130 GA. Hwy 81 SW, Loganville, GA 30249.
Phone: 404 466-3241, FAX: 404 466-9013. Sells via showroom
and mail order.

Omega Electronics
101-D Railroad Street, PO Box 579, Knightdale, NC 27545.
Phone: 919 231-7373, FAX: 919 250-0073. Sells via showroom
and mail order.

Omni Electronics
1007 San Dario, Laredo, TX 78040. Phone: 512 722-5195. Sells via showroom and mail order.

P

P.A.C.E.
1720 West Wetmore Road, Tucson, AZ 85705. Phone: 602 888-3333.

Page-Comm, Inc.
10935 Alder Circle, Dallas, TX 75238. Phone: 214 340-8876.

Paramount Communications Elec.
PO Box 506, Dalton, OH 44618. Phone: 800 431-7777, 216 828-2071, FAX: 216 828-8308.

Parsec Communications, Inc.
13313 Forest Hill Road, Grand Ledge, MI 48837.
    Phone: 517 626-6044. Land Mobile, Microwave and Marine Communications sales and service with amateur service and aircraft communications.

Portland Radio Supply
234 SE Grand Avenue, Portland, OR 97214-1115.
    Phone: 503 233-4904. Sells via showroom and mail order.

Q

Quad Electronics Co.
1420 N. Pace Blvd., Pensacola, FL 32505. Phone: 904 438-3319. Sells via showroom and mail order.

Quement Electronics
1000 S. Bascom Avenue, San Jose, CA 95128.
    Phone: 408 998-5900, FAX: 408 292-9920. Sells via showroom and mail order.

R

R&D Electronics
10511 Phelps Streets, New Orleans, LA 70123.
    Phone: 206 364-8300. Sells via mail order.

R.F. Connection, The
213 N. Frederick Ave., Suite 11, Gaithersburg, MD 20877.
    Phone: 301 840-5477, 800 783-2666, FAX: 301 869-3680. Sells via showroom and mail order. RF connectors and coax.

R.F. Enterprises
HCR Box 43, Merrifield, MN 56465. Phone: 218 765-3254, 800 233-2482, FAX: 218 765-3308. Sells via showroom and mail/telephone order.

RF Microtech
4 Carmel Drive, N. Billerica, MA 01862. Phone: 508 667-8900.

R.F. Products
PO Box 195, Greenfield, IN 46140. Phone: 317 462-6146.

R & L Electronics
1315 Maple Avenue, Hamilton, OH 45011. Phone: 800 221-7735,
513 868-6399, FAX: 513 868-6574. Sells via showroom and
mail order.

Radio Center USA
102 NW Business Park Lane, Kansas City, MO 64150. Phone:
800 821-7323, 816 741-8118. Sells via showroom and mail
order.

Radio Center USA
12 Glen Carran, Circle Sparks, NV 89431. Phone: 800 345-5686,
702 331-7373, FAX: 702 331-3762. Sells via showroom and
mail order.

R & S Electronics Ltd.
(See Canada Listings.)

Radio Comm. of Charleston, Inc.
102 Farm Road, Goose Creek, SC 29445. Phone: 803 553-4101,
FAX: 803 553-3564.

Radio Inc.
1000 S. Main Street, Tulsa, OK 74119.

Radio Place, The
5675A Power Inn Road, Sacramento, CA 95824.
Phone: 916 387-0730, FAX: 916 387-0744. Sells via showroom
and mail order.

Radio Repair By Ed Kuhnley
4484 Tumbleweed Trail, Port Orange, FL 32127.

Radio Works
PO Box 6159, Portsmouth, VA 23703. Phone: 804 484-0140.
Sells via mail order.

RadioKit
169 Jeremy Road, Pelham, NH 03076. Phone: 603 635-2235.
Sells via showroom and mail order.

Reliable Electronics
3306 Cope Street, Anchorage, AL 99503. Phone: 907 561-5515.

Rio Radio Supply, Inc.
515 S. 12th St., Box 1808, McAllen, TX 78501.
Phone: 512 682-5224.

Rivendell Electronics
8 Londonderry Road, Derry, NH 03038. Phone: 603 434-5371.
Sells via showroom and mail order.

Rogus Electronics
250 Meriden-Waterbury Turnpike, Southington, CT 06489.
    Phone: 203 621-2252. Sells via showroom and mail order.

Rosen's Electronics, Inc.
208 Logan Street, Williamson, WV 25661. Phone: 304 235-3677,
    FAX: 304 235-8038. Sells via showroom and mail order.

Ross Distributing, Co.
78 S. State Street, Preston, ID 83263. Phone: 208 852-0830,
    FAX: 208 852-0833. Sells via showroom and mail order.

S

Satellite City
2663 Country Road I, Minneapolis, MN 55112.
    Phone: 800 426-2891, 612 786-4475. Sells via showroom and
    mail order.

Scanner World USA
10 New Scotland Avenue, Albany, NY 12208.
    Phone: 518 436-9606.

Slep Electronics
Highway 441, Franklin South, Otto, NC 28763.
    Phone: 704 524-7519. Sells via showroom and mail order.

Smith Electronics, Dick
PO Box 301 173 E. Broadway, Greenwood, IN 46142.
    Phone: 317 888-7265, 800 872-1373, FAX: 317 888-7368. Sells
    via showroom.

Sound Electronics
103 Arnold Blvd., Lafayette, LA 70506. Phone: 318 984-4090.

Soundnorth Electronics
1802 Highway 53, International Falls, MN 56679.
    Phone: 218 283-9290, 800 932-3337. Sells via showroom and
    mail order. Used gear.

Spokane Radio S.
    25 Girad, Spokane, WA 99212. Phone: 509 928-3073.

Spectronics Inc.
1009 Garfield Street, Oak Park, IL 60304. Phone: 708 848-6777,
    FAX: 708 848-3398. Sells via showroom and mail order.
    Surplus two-way equipment and SW receivers, accessories for
    both.

Surplus Sales of Nebraska
1502 Jones Street, Omaha, NE 68102. Phone: 402 346-4750,
    FAX: 402 346-2939. Sells via showroom and mail order.

T

TNR Technical, Inc.
279 Douglas Avenue, #1112, Altamonte Springs, FL 32714.
Phone: 800 346-0601, FAX: 407 682-4469. Replacement batteries and inserts for popular ham hand-helds; gel cells; custom batteries.

Texas Towers
1108 Summit Avenue, Ste. 4, Plano, TX 75074.
Phone: 800 272-3467, 214 422-7306, FAX: 214 881-0776. Sells via showroom and mail order.

Texpro Sales Canada Inc.
(See Canada Listings.)

Transworld Cable Company
3958 Northlake Blvd., Lake Park, FL 33403.
Phone: 800 442-9333.

Traxit
807 Quince Avenue, Suite #4, McAllen, TX 78501.
Phone: 512 682-6559, FAX: 512 682-1658.

Tucker Surplus
1717 Reserve Street, Garland, TX 75042. Phone: 214 340-0631, 800 527-4642, FAX: 214 348-0367.

U

Universal Radio Inc.
6830 Americana Parkway, Reynoldsburg, OH 43068.
Phone: 614 866-4267, 800 431-3939, FAX: 614 866-2339. Sells via showroom and mail order.

V

VHF Communications
280 Tiffany Avenue, Jamestown, NY 14701.
Phone: 716 664-6345, 800 752-8813, FAX: 716 487-0310. Sells via showroom and mail order.

Valley Radio Center
1522 N. 77 Sunshine Strip, Harlingen, TX 78550.
Phone: 512 423-6407, 800 869-6439, FAX: 512 423-1705. Sells via showroom and mail order.

Van Valzah Company, H.C.
1140 Hickory Trail, Downers Grove, IL 60515. Phone: 708 852-0472, Orders: 800 HAM-0073, FAX: 708 852-1469. Sells via mail order.

W

Western Radio Electronics
4797 Ruffner Street, San Diego, CA 92111. Phone: 619 268-4400, 800 777-4973, FAX: 619 279-7048.

Williams Radio Sales
600 Lakedale Road, Colfax, NC 27235. Phone: 919 993-5881.
Sells via showroom, hamfests and mail order.

The Wireman, Inc.
261 Pittman Road, Landrum, SC 29356. Phone: 800 727-WIRE,
803 895-4195, FAX: 803 895-5811. Sells via mail order.

Woody's Antenna & Tower Service
PO Box 222, Levittown, NY 11756. Phone: 516 731-6662. Sells
via mail order.

Canada

ALL Band Radio Products
3378 Douglas Street, Victoria, BC V8L 3L3.
Phone: 604 477-1829, 604 477-9665. Sells via mail order.

Atlantic Ham Radio, Ltd.
368 Wilson Avenue, Downsview, ONT M3H 1S9.
Phone: 416 636-3636, FAX: 416 631-0747. Sells via showroom
and mail order.

Century 21 Communication
23 McLeary Court, Unit 23, Concord, ONT L4K3R6 (Metro
Toronto). Phone: 416 738-0000, FAX: 416 738-1169. Sells via
showroom, mail order.

Com-West Radio Systems, Ltd.
8179 Main Street, Vancouver, BC V5X 3LZ.
Phone: 604 321-1833, FAX: 604 321-6560. Sells via showroom
and mail order, catalog available ($2.00).

Hobbytronique Inc.
8104-A Trans Canada Highway, Ville St. Laurent,
Quebec H4S 1M5. Phone: 514 336-2423, 800 363-0930,
FAX: 514 336-5929.

MacFarlane Electronic, Ltd., H.C.
R.R. #2, Battersea, ONT K0H 1H0.
Phone: 613 353-2800, FAX: 613 353-1294. Sells via showroom
and mail order.

R & S Electronics Ltd.
157 Main Street, Dartmouth, Nova Scotia B2X 1S1.
Phone: 902 434-5235, FAX: 902 434-5590. Sells via showroom
and mail order.

Texpro Sales Canada Inc.
5035 North Service Road, Unit D16, Burlington, ONT L7L.5V2,
Canada. Phone: 416 332-5944, FAX: 416 332-5946. Sells via
showroom and mail order.

Glossary

AM: Once a common voice mode on the ham bands, AM (short for amplitude modulation) has been largely replaced by a variation, SSB.

Amateur service: A radio service whose purposes are self-training, intercommunication and technical investigations. Amateur Radio operators conduct these activities on a noncommercial basis. The FCC regulates the amateur service in the U.S.

Amateur TV (ATV): A mode of communication that involves sending television signals to other hams.

American Radio Relay League (ARRL): The nationwide organization of ham radio operators in the US.

AMSAT: The Radio Amateur Satellite Corporation, an international organization that designs, builds and promotes use of ham radio satellites.

AMTOR: AMateur Teleprinting Over Radio, a mode of digital communication that provides error detection.

Autopatch: A means of connecting a ham rig to the telephone lines.

Band opening: A condition of the ionosphere that permits bands of radio frequencies to propagate between two points.

Base transceiver: A radio designed to be used at a home station.

Beam antenna: An antenna that radiates (and receives) better in one direction than others.

Code (see **Morse code**)

Contact (also see **QSO**): A two-way conversation between two hams.

Contesting: A popular ham activity that involves making as many contacts as possible during a set amount of time. Various organizations and groups, such as ARRL and *CQ,* sponsor contests.

CQ: An abbreviation for a CW call to any other station.

CQ: A monthly ham radio magazine.

CW (see **Morse code**)

Cycle: A complete wave—one that rises to a maximum point, falls past the starting point to a minimum point, and then rises back to the starting point.

Dipole Antenna (also see **Half-wave dipole**): A wire antenna having two legs of equal length.

DXing: Trying to communicate with distant stations.

DXpedition: A trip to a place that has little or no ham radio activity. Hams go on DXpeditions to provide contacts with a "rare" country.

Elmer: An experienced ham who helps a beginner earn a license and get on the air.

FCC: The Federal Communications Commission, which regulates the amateur service in the US.

Feed line: The wires or cable used to connect a transceiver to an antenna.

Field Day: Hams set up and operate stations away from electrical power, often in remote locations, during the fourth full weekend in June. In this way, Amateur Radio operators practice communicating under emergency conditions.

FM: The most popular VHF mode, FM stands for frequency modulation. VHF repeaters use FM because it is less noisy than other voice modes.

Frequency (also see **Wavelength**): The rate at which a wave goes through a complete cycle. Radio-frequency energy at 28.3 MHz means the wave has a frequency of 28,300,000 cycles per second. When the frequency is large, wavelength is small, and vice versa.

Half-wave dipole antenna (also see **Dipole antenna**): A type of wire antenna that has two legs of equal length and that is fed in the center. Its total length is a half wavelength long at the desired operating frequency.

Ham: Another name for an Amateur Radio operator.

Hamfest: An organized gathering of ham radio operators. One of the most popular parts of a hamfest is the flea market.

Hand-held transceiver: Small, portable radio popular with beginners. Can be carried almost anyplace.

Hertz: (also see **kHz** and **MHz**): Another name for cycles per second—the rate at which a wave completes a full cycle.

HF: High-frequency radio waves, extending from 3 to 30 MHz.

Ionosphere: A layer of the atmosphere responsible for long-distance radio communications.

kHz: The abbreviation for kilohertz, or thousands of cycles per second.

Line of sight: Some radio signals normally travel point to point. VHF, UHF and microwave signals are usually line of sight: If you can see the point you're trying to communicate with, you will be able to reach it.

Mag-mount antenna: A mobile antenna that attaches to a vehicle with a magnet for easy removal.

Meteor scatter: A mode of communication that involves bouncing a radio signal off a meteor trail.

MHz: The abbreviation for megahertz, or millions of cycles per second.

Microwaves: The frequencies from 1000 MHz (1 GHz) to 250 GHz.

Mobile transceiver: A radio designed to be used in a vehicle. Many mobile radios are made for one or more VHF and UHF bands, but some HF radios are designed for mobile use as well.

Mode: A type of communication, such as AM, FM, SSB and CW. Hams can use any of several different modes, depending on class of license and band.

Moonbounce: Bouncing VHF, UHF and microwave radio signals off the moon. Requires very high power and large arrays of Yagi antennas or large dish antennas.

Morse code: The system of long and short sounds that stand for letters of the alphabet, numbers and punctuation marks. Most hams around the world know and use Morse code. Often abbreviated CW.

Multimode: A transceiver that allows more than one type of communication. Many HF transceivers can send and receive CW, SSB and AM, for example.

Multimode controller: A device used in digital communications such as packet and AMTOR that connects between the computer and the transceiver.

Novice: The class of license that permits beginners to operate CW on part of four HF bands and the 222- and 1296-MHz bands. To earn this license, you must pass a code test and a written test.

Omnidirectional: An antenna that radiates (and receives) equally well in all directions.

OSCAR (see **Satellites, amateur**)

Packet radio: A type of computer-to-computer communications used to send and receive error-free messages.

Propagation: The study of how radio waves travel.

Q signal: A three-letter abbreviation beginning with the letter Q. Used in Morse code communications to save sending time.

QSL: A confirmation of a two-way conversation between two hams.

QSL card: A postcard that serves as a written confirmation of an on-the-air contact. Many hams collect QSL cards from around the world.

QSO: A two-way Amateur Radio contact.

QST: The monthly journal of the ARRL.

Radio-frequency energy (RF): Waves whose frequencies range from approximately 20 kHz (20,000 cycles per second) to 300 GHz (300,000,000,000 cycles per second).

Repeater: Automatic relay station. Usually installed on a hill or tall building to increase communications range.

Rig: Another name for transceiver.

RTTY: Radioteletype, a mode of digital communication that involves sending messages on a keyboard. These messages are received on a printer, video terminal or personal computer screen.

Rubber duck: A small flexible antenna that comes with most hand-held transceivers.

SAREX: A space shuttle mission that includes ham radio. Stands for Shuttle Amateur Radio EXperiment.

Satellites, amateur: Satellites designed and built by hams for use by hams. Many amateur satellites are called OSCARs, for *O*rbiting *S*atellites *C*arrying *A*mateur *R*adio. Amateur satellites act as relay stations that enable hams to communicate over long distances using VHF/UHF bands.

Sideband (see **SSB**)

Simplex: A VHF or UHF contact made directly, without the use of a repeater.

Single sideband (see **SSB**)

Slow-scan TV (SSTV): A mode of communication that involves sending still pictures to other hams.

SSB: Single sideband, a voice mode used for long-distance communication.

Sunspot cycle: The 11-year period during which sunspots, dark patches on the sun, appear and disappear in a predictable way. The number of sunspots affects how HF radio waves travel on earth.

Technician: The class of license that permits beginners to use all authorized modes on all amateur frequencies above 30 MHz. Technicians who pass a code test earn Novice HF privileges as well.

Terminal node controller (TNC): A device used in packet radio that connects between the computer and the transceiver.

Transceiver (also see **Rig, Hand-held transceiver, Mobile transceiver, Base transceiver**): A radio that sends and receives radio signals.

UHF: Ultra-high frequency radio waves, extending from 300 to 1000 MHz.

Vertical antenna: An omnidirectional antenna used most often for mobile operation, but also used for home (base) stations.

VHF: Very-high frequency radio waves, extending from 30 to 300 MHz.

W1AW: The Headquarters station of the American Radio Relay League, in Newington, Connecticut. W1AW sends out news bulletins and Morse code practice.

Wavelength (also see **Frequency**): The distance a radio wave travels during one cycle. When wavelength is large, frequency is small, and vice versa.

Wire antenna: An antenna popular with beginners because it works well, is inexpensive and is easy to install. Some are designed to work on several HF bands.

Worked All States (WAS): An award sponsored by the ARRL for contacting hams in all 50 US states.

Worked All Zones (WAZ): An award sponsored by *CQ Magazine* for contacting stations in the 40 zones of the world, as defined by CQ.

Yagi: A popular type of beam antenna. Increases transmitted and received signal strength by concentrating its sensitivity in one direction.

Index

NOTE: G-1 means Glossary page 1, and so on.